TOUCH MATTERS

TOUCH MATTERS

HANDSHAKES, HUGS, AND THE NEW SCIENCE ON HOW TOUCH CAN ENHANCE YOUR WELL-BEING

MICHAEL BANISSY

CHRONICLE PRISM

First published in the United States of America in 2023 by Chronicle Books LLC.
Originally published in the United Kingdom in 2023 under the title *When We Touch* by Orion Spring, an imprint of The Orion Publishing Group Ltd.

Copyright © 2023 by Michael Banissy.
All rights reserved. No part of this book may be reproduced in any form without written permission from the publisher.

Library of Congress Cataloging-in-Publication Data available.

ISBN 978-1-7972-2144-1

Manufactured in the United States of America.

Typeset by Input Data Services, Bridgwater, Somerset.
Typeset in Duotone, Industry Inc, and Warnock Pro.

This book contains advice and information relating to health and interpersonal well-being. It is not intended to replace medical or psychotherapeutic advice and should be used to supplement rather than replace any needed care by your doctor or mental health professional. While all efforts have been made to ensure accuracy of the information contained in this book as of date of publication, the publisher and the author are not responsible for any adverse effects or consequences that may occur as a result of applying the methods suggested in this book.

This book contains anecdotes and quotes. To maintain confidentiality, names and other data have been modified in certain instances.

10 9 8 7 6 5 4 3 2 1

Chronicle books and gifts are available at special quantity discounts to corporations, professional associations, literacy programs, and other organizations. For details and discount information, please contact our premiums department at corporatesales@chroniclebooks.com or at 1-800-759-0190.

CHRONICLE BOOKS
SAN FRANCISCO

Chronicle Prism is an imprint of Chronicle Books LLC,
680 Second Street, San Francisco, California 94107
www.chronicleprism.com

CONTENTS

Introduction   vii

PART 1   WHY WE TOUCH

1. Developmental Touch: The Origins of Our Most Underappreciated Sense   3
2. Scientists Who Stroke: The Neuroscience of Touch   19

PART 2   GOOD TOUCHING

3. Healthy Touch: Can Hugs Conquer the Common Cold?   41
4. Touch Hunger: What Happens When We Don't Receive Enough Touch?   63
5. Tactile Intimacy: From Sex to Spooning, the Role of Intimate Touch in Relationships   87

PART 3   TOUCH TRAITS

6. Touchy-Feely or Avoid at All Costs: Our Touch Personas   111
7. Touch Culture: How Our Backgrounds Affect How We Perceive Touch   137

PART 4 TOUCH MATTERS

8. Social Touch: The Hidden Secret to Effective Teamwork — 153
9. Do Touch, Don't Touch: The Murky World of Touch at Work — 175
10. Digital Touch: The Future of Touch in Our Society — 195

Conclusion — 217
Appendix: Further Materials — 221
Acknowledgments — 223
Endnotes — 225

INTRODUCTION

What does touch mean to you?

On the face of it, this is a relatively simple question. Yet with a sense so intimately linked to our daily lives, the answer is far from straightforward. Under the term fall the most intimate and the most formal of behaviors—a caress, a hug, a handshake. Touch can bring pleasure. Touch can bring pain. Touch can trigger a complexity of emotions and memories.

Many might be able to recall a time when the experience of touch brought them closer to other people. What made that experience special? What was it about the quality of tactile communication that helped you connect with someone else? Caring physical contact, like hugging or hand-holding, is vital to convey reassurance, empathy, and affection. These touches act as a social glue, helping to form and reinforce bonds between us.

The relationships that touch facilitates are essential, because social connections are vital for our health and well-being. The proven benefits of strong social connections include increased longevity, a strengthened immune system, and positive mental health. One landmark study found that people with stronger social relationships have a 50 percent increased likelihood of survival compared to those with weaker social connections.[1] Touch is a core ingredient in social contact, helping to generate a positive loop for social, emotional, and physical well-being to thrive.

The benefits of touch on physical and mental health extend further. For instance, as we will learn later in this book, even a short and gentle touch has been found to lower anxiety, reduce stress, and decrease symptoms of depression in the person who is touched.

Many of us might be able to recall a time when a simple comforting touch made a difference. Can you think of a time when you consoled someone? Or perhaps a time when touch helped calm you before a stressful event like having an operation? In some cases, the support provided by touch can be more of an aid to help people overcome adverse events than words alone.

As a social neuroscientist (someone who studies the brain basis of social interaction), I have been fascinated by touch for many years. In my work, I've sought to understand why we touch others and what happens when we do. What drives differences between us in how we think and feel about touch? What is it about the quality of tactile communication that so positively impacts our relationships, happiness, and health?

From working in the laboratory to observing people in the real world, I've tried to get a handle on the meaning of touch from the earliest stages of life to our final moments. I've tried to understand what makes touch so fundamental to human connection and existence.

Through this work, I've been lucky to investigate touch with people from various backgrounds and in different settings. Some examples include the role of touch in the workplace, in theatre and dance, in healthcare, and across our daily relationships. I've worked closely with organizations to try to find ways to balance the positives that come from appropriate and supportive touch with the importance of keeping consent at the heart of our interactions. And with those seeking to develop new approaches to ensure that even if we lack touch in our lives, we can still have access to opportunities to experience the positive outcomes that supportive touch can bring.

Away from my work as a scientist, touch is an integral part of my life, as it is for many of us. It is a sense that has often brought me comfort. A sense that has helped me to share a connection with others. A sense that has helped me explore and understand the world around me.

While touch can be positive, there is no doubt that it can also be challenging. The words I've written above will not resonate with everyone. I fully acknowledge that I am privileged to have had a positive relationship with touch throughout my life. Many have not.

Sadly, touch is a sense that has often been misused. Nowadays, you do not have to search far to find examples of exploitation or abuse through touch. Egregious examples of inappropriate touch and sexual harassment are distressingly common. The volume of reports from across the world of people abusing their licence to touch others has been a troubling feature of recent times. Considering this, I can understand why some people have argued that we should place greater scrutiny on the impact of touch and consent.

We know that touch affects us all, but its meaning is often complicated and nuanced. What might be a meaningless touch to some may be significant to others. Our thoughts about touch are often subconscious and connected to our personal history of tactile experiences.

This is true from person to person but can even be the case with the same person from situation to situation. Can you imagine an example where you might feel comfortable with touch in one setting but not another? If your hairdresser touched your arm or shoulder while in the salon, it might be okay, but what if they stroked your arm to greet you in the coffee shop? Context matters, and we'll explore this in more detail later.

We rely on touch every day of our lives. It makes us who we are. It helps us connect. Despite this, many feel out of touch with the world. Studies suggest that over half of us long for more touch in our lives. But people increasingly report a reluctance to touch.

When we touch, we must navigate a delicate landscape of past, present, and future encounters. How do we balance these competing factors? How can we ensure that our tactile behaviors are most appropriate for the individual needs of each person we interact with? Maybe you've thought about these questions yourself. Perhaps you have not.

The reality is that touch is a sense that impacts many parts of our lives in unexpected ways. Yet we seldom stop to consider our everyday tactile interactions. Until recently, that is.

On January 21, 2020, I woke up in a hotel room in the heart of MediaCity, a 200-acre development beside the Manchester Ship Canal in Salford. Historically known as one of the world's most extensive river navigation canals, the Manchester Ship Canal is heralded as a feat of Victorian engineering that connected the city to global trade routes.

Today, its banks are somewhat different. They are home to some of the UK's world-leading media organizations, including the British Broadcasting Corporation—or the BBC, as it is more commonly referred to.

I had a big day ahead of me: a string of radio and television interviews throughout the day until a final appearance on the BBC evening news. Why? Touch. We were launching what would become one of the world's largest contemporary studies on people's attitudes to and experiences of touch around the globe. A study called the Touch Test.

That morning in January 2020, we had no idea what was coming our way.

On the very same day as our launch, the first confirmed case of coronavirus (COVID-19) was identified in Washington State. The previous day, epidemiologists from the Chinese Center for Disease Control and Prevention published an article indicating that the first cluster of patients with "pneumonia of an unknown cause" had been identified on December 21, 2019.[2]

At the end of January 2020, the director general of the World Health Organization (WHO) declared a public health emergency. In the following weeks, the stark reality of the scale of this emergency became clearer. The number of world cases surpassed 100,000 in March 2020, then 1 million in April.

At the time of writing, in spring 2022, the current global number of cases stands at 521 million, with 6.26 million deaths. New variants have developed, and cases continue to rise. COVID-19 has changed the world.

COVID-19 was also a game-changer for how we touched. As societies understood that the virus could spread through close contact, various restrictions and new social norms developed. National lockdowns were introduced, with strict rules limiting interactions with others outside our households and ordering people to stay home. Public health messages advised regularly sanitizing hands and urged people not to touch their faces. We were warned to keep our distance from each other and to be cautious of touch.

As a touch researcher, I always thought I understood the importance of touch in our lives. Yet like many people, it was not until touch was taken away that I realized just how important it was to me. For many, the past few years have put touch into focus. This forgotten sense has returned to the public eye.

When we launched our global Touch Test study, we did not know how many people would complete it. But even though COVID-19 took hold during our testing period, close to 40,000 people living in 112 different countries took part. The timing could not be controlled—the test was even referred to as "one of the most ill-timed scientific surveys carried out"[3]—yet it turned out to be a fascinating window into what touch meant to different people across the globe at a time when they were deprived of it more than ever before.

Many people also reached out during and after broadcast programs linked to the survey results. They shared their feelings about a lack of touch during pandemic restrictions. And they shared reports of how it felt to have the first hug from a loved one post-lockdown.

The diversity and richness of these experiences fuelled my curiosity to dig deeper into my understanding of touch. *Touch Matters* shares these scientific and personal insights. It is a journey into the new and emerging science behind our everyday tactile experiences.

I cannot claim to be able to tell every story of touch. Whole volumes have been, and should continue to be, written on some of the darker sides of the subject—the unacceptable nonconsensual forms of touch captured by social movements like #MeToo.[4] I would never wish to minimize or omit these stories from a discussion on touch. However, I acknowledge that coming from a place of touch privilege, I could never do justice to them.

Instead, *Touch Matters* focuses predominantly on the daily interactions involving touch that we consider appropriate and consensual, the subtleties of common occurrences of touch that many will see in their lives from one day to the next.

My aim is not to argue in favor of or against touch between individuals. It is instead to reflect on the nuanced nature of human touch. I will explain the fundamental principles of why we touch and its consequences. We'll delve into everything from everyday tactile interactions (like hugs, handshakes, and high-fives) to how we can help people experiencing touch starvation. We will seek to understand what touch means, how it defines us, and how it contributes to health and well-being. I will try to answer questions offering the potential to help us communicate better via touch.

While this is by no means a complete handbook of touch in the modern world, I hope it will give you a grounding in the subject and leave you with a greater understanding of the importance of everyday touch for yourself and those around you.

For now, I want to return to the question with which we opened this chapter: What does touch mean to you?

I was hoping you could take a moment to think of three words that might best describe the meaning of touch in your life. You can write them down, keep them in mind, whatever works best.

When my team asked this question of nearly 40,000 people worldwide in 2020, the three words most commonly used to describe what touch meant were: "comforting," "warm," and "love."[5] These words were similar across all regions, with the occasional inclusion of words like "caring," "connection," and "affection." They were similar among male, female, and gender nonbinary adults. They were similar for those with and without a range of health conditions. You may also be in this majority.

I was surprised by the consistency of the responses. We often think of touch as being a fluid sense. A sense that can fluctuate based on a diverse range of experiences and preferences. This is true, as we will come to see. Yet, at its core, touch is often identified by many people as a sense that comforts and connects us. Why would that be? To answer this question, let's start our journey into the world of touch by looking at the humble beginnings and biology of this underestimated sense. An origin story that begins from our earliest moments.

PART 1

WHY WE TOUCH

CHAPTER 1

Developmental Touch: The Origins of Our Most Underappreciated Sense

On May 29, 2021, an American mother made headlines with a ten-second clip posted on TikTok. She gained over 20 million views with her video of a toy fish gently patting her baby's bum until the baby fell asleep.

While it might sound funny, believe it or not, the sleeping fish is a product marketed to parents. The rhythmic patting of the toy is thought to mimic the calming touch of a caregiver. Many parents across social media have reported that the fish helps their baby, and them, get a better night's sleep.

To be clear, I am not here to comment on the sleeping fish from a scientific perspective. But having been in that fretful situation where you find yourself slowing down the speed of comforting touch while gradually moving away so that the child doesn't notice you sneaking out after soothing them to sleep—well, I can see the appeal. If only I had known of such a toy back in the day!

All joking aside, did you know that touch is one of the first senses we acquire? In fact, touch processing can develop early in the prenatal period between conception and birth.[1] For instance, during pregnancy, it has been shown that the fetus responds to maternal touch on its mother's abdomen with tactile exploration. It reaches out and touches the uterus wall.

On the one hand, you might think this would make us experts in touch—we have, after all, been using touch to connect with the world from our earliest moments. But I'm sure I'm not alone in recognizing that life teaches us that touch is complicated, whether due to a pandemic, social norms, past experiences, or simply difficulty understanding what touch means from one person to another. In adulthood, many can find touch awkward at best.

There are many reasons for this, which we will address throughout the book. But for now, if we want to truly understand a sense with us from so early in life, we need to retrace our steps a little. We need to look at how touch begins and contributes to our development.

A GENTLE CARESS

Before we dive into the science, it's probably important for me to note that the opening chapters of this book contain a bit more about the biology of touch than the others. We also begin with more of a focus on babies than you will see throughout the rest of the book: This is because much of what we know about touch during development comes from studies on young infants.

If you want to learn about touch but aren't too worried about all the cells and brain regions that play a role, then don't worry; the book is set up in such a way that you don't need to know all the biology to get the most out of the chapters that follow. There are key takeaway sections in each chapter so that you can move between sections in a way that works for you. On the other hand, if you have even a slight interest in the biology of touch or the importance of the touch that we experience during development, what follows will appeal to you.

To get us started, let me circle back to my own childhood.

When I was a teenager, my life changed substantially for the better. At thirteen, my younger sister joined the family. I recall several people noting with curiosity the age gap between us. Many still do. The age gap was a great thing for me, giving me a different type of relationship with my sister that has endured to this day.

Being that bit older allowed me to be part of the caregiving process. I was able to be a role model and pass on things I'd learned. Okay, I may have moaned and groaned at times—I was still a teenage boy, after all. But I never complained about taking care of my sister.

Some of my fondest memories involve making her laugh and smile from the earliest days of her life. I recall the beaming smiles I would get when she was a baby by gently stroking her cheeks or tiny arms. In hindsight, perhaps positive experiences like these led me to become a scientist who wanted to study touch.

My choice to stroke my sister's arms aligns with research on how caregivers tend to touch newborn babies. In 2019, researchers from Concordia University examined how mothers and fathers touched their babies during the first hour after birth. Both caregivers displayed similar types of touch when interacting with their infants for the first time. Stroking and caressing were the forms of touching engaged in most often by both mothers and fathers. Conversely, kissing was the form of touch that was used the least, by both mothers and fathers, during these first-hour tactile interactions.[2]

Our tendency to gently stroke infants may not merely be a random act. A type of nerve fiber in the skin appears particularly tuned to comforting and caregiving touch, like gentle skin stroking. These nerve fibers are called C-tactile afferents or C-tactile fibers (CT fibers). They respond best to things like gentle skin stroking.

Scientists refer to touch that activates these CT fibers as CT-optimal touch. For our purposes, the best way to picture this is to imagine a touch that resembles the type of gentle caress you might give to a loved one: slow, gentle skin stroking at a speed at which you might intuitively stroke a baby.

In fact, CT-optimal touch appears to be intuitive to many people.[3] If you watch caregivers spontaneously stroking their babies, you will see that they often naturally do this at the optimal speed for CT fibers to respond.

There is some variability in this; for instance, a recent study on infants found that stroking speed is related to a mother's heartbeat before stroking. That is to say, the mother's prior state impacted stroking speeds—if their heart rate was higher, they would stroke their child faster, and vice versa. Still, even adults will typically spontaneously stroke each other at speeds that target CT fibers when freely interacting with people with whom they have close relationships.

Is there a reason parents intuitively engage in this type of stroking with their children? An exciting body of new research suggests that the answer might be yes.

In 2018, a study led by researchers from the University of Oxford examined brain and behavioral responses to a painful medical procedure in newborn babies—a pinprick blood test used commonly as part of a mixture of tests to detect genetic conditions.[4]

The researchers divided the babies into two groups. One group was gently stroked with a soft brush at an optimal speed for CT touch before experiencing the pinprick test. The other group did not experience touch before testing.

Gentle stroking before experiencing pain led to reduced pain-related brain activity and behaviors like withdrawal reflex. In other words, caring touch helped infants get through exposure to a painful experience like a pinprick to their heel.

The benefits of CT touch to infants are not just physiological. They extend to social and emotional functioning as well. It has been found that babies can appear happier after being gently stroked.

Infants can also differ in how they perceive the social signals of other people when they are stroked. In 2021, researchers from the University of Milan-Bicocca explored how seven-month-old babies responded to emotional faces when stroked or squeezed by their mothers. When infants see angry faces, they typically show signs of avoiding them by looking away. In this experiment, stroked babies showed less avoidance of an angry expression. In contrast, babies who were gently squeezed rather than stroked still looked away.[5] It seems that gentle stroking helped the babies be more confident in exploring the social world around them, even when it seemed scary. CT touch influenced how infants responded to the emotions of others.

In fact, throughout childhood, there is evidence that children who share supportive touch from caregivers tend to be more social with others. This kind of touch doesn't just need to be stroking. Researchers from the University of Notre Dame have shown that children who are hugged more often when upset show greater concern and care towards others.[6]

It also turns out that some of the first touch sensations we experience in the world can profoundly impact social development throughout life, even decades after we first share them. In 2021, researchers from Reichman University reported a study that took them almost 20 years to complete.[7] They meticulously followed newborn babies and their mothers from birth to adulthood to study social behaviors and brain responses in newborns who received varying degrees of initial maternal tactile contact shortly after birth.

Three groups of infants were tested. One group was full term and had tactile contact with their mothers when they were born. Another group, born preterm, engaged in skin-to-skin care interventions with their mothers. And finally, a group born preterm was cared for in incubators without touch from their mothers. More succinctly, the first two groups had very early maternal tactile contact as newborns, but the final incubator group did not.

Over two decades, the researchers videotaped mother–child interactions in the homes of the three groups in infancy, preschool, adolescence, and adulthood. What they found may astonish you. Differences in early tactile experiences as newborns had a demonstrable impact on social processing some 20 years later.

The infants who received more touch from their mothers after birth showed more coordinated social behaviors with their mothers across development. In other words, the mothers and children who shared touch early were more in sync when interacting socially. This was true for infants born full term with tactile contact and infants born preterm who engaged in skin-to-skin care interventions.

The impact of early tactile interaction on social responses did not stop there. One of the most intriguing parts of the Reichman University study was that the researchers did not only study mother–child interactions. They also investigated what happened in adulthood when the now grown-up babies observed social signals from other people.

This was achieved by studying the brain basis of affective empathy. In this task, the participants (now adults in their twenties) had their brain activity recorded as they watched videos of strangers experiencing joy, distress, or sadness in social contexts.

People with more mother–child synchrony showed greater sensitivity to the emotional stories of other people in the amygdala and insula—both core structures of the social brain. The enhanced social interaction between mother and child, connected to early touch, predicted social processing towards other people.

In a nutshell, early tactile interactions contributed to empathy responses towards strangers two decades later. A staggering demonstration of the powerful impact that touch throughout development can have on our social world.

THE CASE OF THE TACTILE BABOONS

At this point, although it's becoming apparent that close physical contact with a caregiver during development can affect social processing, you might be wondering *why*.

Clues to help us answer this can be drawn from an unlikely source—some of our primate relatives: monkeys and apes. Although we may not immediately think of mammals like monkeys and apes as being family, they display very similar DNA to that of humans. We share nearly 99 percent of our DNA with some of our closest animal relatives—the bonobo and the chimpanzee. Scientists use the similarities to make meaningful comparisons between people and other mammals. They use these comparisons to address questions like why do we touch one another?

One tactile behavior shared across different primates is using touch to groom. We see this behavior as plucking—pulling—and rhythmic tactile movements across the body. If you've ever watched a nature documentary featuring monkeys, you will have seen that grooming—and, by proxy, touching—is widespread in primates. Some have been reported to spend nearly 20 percent of their day in mutual grooming!

But why do primates spend so much of their time engaging in touch?

Intuitively, we might immediately think about grooming as something to do with personal hygiene. You might imagine that a larger monkey will need to spend more time grooming than its smaller cousins simply because it has more fur to be groomed. But studies show that is not the case.

Many primates indeed do self-groom to stay clean and healthy. Yet some species can remain hygienic even with less than 1 percent of their time spent on grooming.

The time that primates spend grooming is much more than the amount necessary simply to keep their fur clean. Research shows that several monkeys and great apes groom each other for reasons beyond hygiene. What could these be?

It turns out that for many primates, grooming is an incredibly social affair. All the primate species I've just mentioned groom together in groups. The time spent grooming relates not to the size of the primate, but the size of the primate's social group—the more primates in a group, the more time the group spends grooming each other. It is clear then that grooming has something to do with social relations between partners within a primate group.

Social grooming in primates is also incredibly consistent. The partners that groom together on one occasion often do so repeatedly. Some have even shown that this consistency between grooming partners can last for years. Those rhythmic tactile movements that take place when two monkeys groom could be part of forming a powerful social connection.

Observations like these have led anthropologists and evolutionary psychologists to suggest that touch during social grooming is all about bonding.[8] Researchers, such as Robin Dunbar of the University of Oxford, contend that one of the reasons why primates socially groom each other is to build relationships that can help them throughout life.

Take the gelada baboon. The likelihood of a female gelada baboon assisting another female baboon under attack is related to the time they have spent grooming each other. A female is more likely to come to the aid of another female with whom she shares a close grooming relationship. At the same time, those who don't share that grooming bond are less likely to intervene. The touch that occurs during social grooming builds relationships and alliances that help in other situations.

To bring this back to humans, these primate studies imply that touch can contribute to developing cooperative social relationships. Given this, it is perhaps not surprising that touch during human development relates to brain regions and behaviors involved in processing information about our social world.

NO BABY UNHUGGED

By now, we are building a picture of the short- and long-term impacts of early caregiving touch on social interaction. But do they extend to other aspects of life too?

Forms of touch—like massage and being held with skin-to-skin contact—have also been shown to support physiological markers of health and well-being from birth.[9] In some studies, preterm babies who received these forms of caregiving touch gained 47 percent more weight and were released from the hospital five days earlier than preterm babies who were touched less.

The benefits of skin-to-skin touch in early life can continue to have an impact even several years after birth. In one study, premature babies who experienced skin-to-skin contact were compared to babies who received incubator care without that direct contact. At ten years of age, those who had received skin-to-skin contact as preterm babies had better sleep patterns, improved physical responses to stress, more advanced autonomic nervous systems, and better ability to adapt their thinking than those who were in incubators. Research findings like this are part of the reason why the World Health Organization now recommends skin-to-skin care for babies weighing 4.4 pounds or less at birth.[10]

In fact, in general, caregiving touch can profoundly impact healthy development. In my quest to understand the importance of early caregiver contact, I encountered the most provocative scientific validation of its significance that I could imagine. This came from the field of epigenetics.

Epigenetics is an area of study that considers how environmental influences can contribute to how genes work. Epigenetic factors don't necessarily change our DNA. Instead, they impact how our body reads a DNA sequence. They do this by regulating behaviors like whether genes are turned on or off.

There are now early signs suggesting that caregiving touch can potentially trigger a form of epigenetic protection to brain development.[11] One study, led by researchers from the University of British Columbia, asked mothers to complete a diary reporting caregiving behavior. The researchers calculated the hours each mother spent engaging in tactile behaviors like holding or carrying infants. The results showed that high levels of maternal contact were linked to differences in genomic regions that play a role in immune and metabolic functioning. In other words, motherly touch impacted how genetic differences in immunity and metabolism were expressed. This could have lasting contributions to our health.

Impacts on sleep, feeding, responses to stressful situations, social behaviors, and even our genetics. I was sold. Early caregiving touch matters!

I was also a little sad. I began to think about situations where sometimes early contact between newborn infants and their parents simply cannot happen. For instance, preterm infants who can only be cared for in incubators. Some births can be complicated and require the mother to be withdrawn for her care and survival. What are organizations doing to ensure that there is still appropriate touch available to infants in situations like this?

The answer is quite a lot.

As data on the importance of early touch has built, neonatal care has been redesigned to facilitate as much uninterrupted contact as possible between caregiver and baby. Where this is not possible, others have stepped in with initiatives that might help give babies the much-needed tactile contact they crave.

While writing this book, I learned about Huggies' No Baby Unhugged campaign.[12] This incredible campaign aims to ensure that babies get skin-to-skin hugs even if their parents cannot provide them.

Much of the campaign's work has focused on supporting hospitals across North America to establish volunteer hugging and cuddling programs. If parents of newborns in intensive care units cannot hold their children after birth, trained volunteers can step in to help give babies touch at this vital time of their lives. These volunteer hugs are a powerful low-cost intervention that could have many beneficial effects on babies in hospitals.

Alongside these attempts, the organization UNICEF has been leading the charge to ensure that skin-to-skin contact should be part of their Baby Friendly standards worldwide.[13] The Baby Friendly standards are a road map to help transform care for "all babies, their mothers, and families." According to UNICEF UK, "Skin-to-skin contact is a key part of the UNICEF UK Baby Friendly Initiative standards."

UNICEF argues that skin-to-skin contact should be valued and supported in hospitals worldwide. Their guidance includes ensuring that all mothers "have skin-to-skin contact with their baby after birth, at least until after the first feed and for as long as they wish," and that "mothers and babies who are unable to have skin contact immediately after birth are encouraged to commence skin contact as soon as they are able, whenever or wherever that may be."

They also suggest that neonatal units make sure that "parents have a conversation with an appropriate member of staff as soon as possible about the importance of touch, comfort, and communication for their baby's health and development."

Many neonatal units now do this.

*

Some organizations have also turned to touch to help parents too: Although society often celebrates birth as a happy event, there is no doubt that it is a very stressful and sometimes traumatic experience. Many new parents can experience changes in mental and physical health afterward. Recent research suggests that early skin-to-skin contact with their newborn child may help some women with postpartum stress and depression symptoms.

One study that speaks to this is a 2019 randomized control trial conducted by the Society for Applied Studies, a research institute in New Delhi. The trial included just under 2,000 mothers of low-birth-weight infants born between April 2017 and March 2018. The mothers were either supported in practicing kangaroo care techniques (methods of holding a baby skin-to-skin) or received standard care that involved home visits without extra kangaroo care support. Engaging in kangaroo care reduced depression symptoms for mothers, leading the authors to conclude that it might cut the risk of moderate to severe maternal postpartum depression.[14]

It's not just mothers who benefit from skin-to-skin; a study published in 2019 by researchers at Nanfang Hospital of Southern Medical University showed that new fathers exhibit lower anxiety and depression following skin-to-skin contact with their new baby.[15] In this work, newborn babies were divided into two groups. One received skin-to-skin contact from their fathers shortly after a Caesarean birth. The other did not. The babies in the skin-to-skin group cried less, ate sooner, and had a more stable heart rate and forehead temperature than the babies who didn't receive this contact. The fathers in the skin-to-skin group had lower self-reported anxiety and depression levels than the other fathers. To put it simply, touch helped both father and baby.

An important implication of this field of research is that it is not just the birthing parent and the baby who can gain from close tactile contact in early life. Although the research is just emerging, and is often restricted to fathers, one would suspect that these benefits would be seen in other non-birthing caregivers involved in early infant interactions. This is good news for the many families that don't always revolve around a mom and a dad. It also offers a positive message to initiatives like No Baby Unhugged, since it implies that babies may gain from close contact with caregiving volunteers and professionals.

Putting everything together, these results build a convincing picture showing that touch is a critical ingredient in allowing babies and their caregivers to thrive. And medicine is adapting to this emerging evidence, with benefits for parents and babies alike.

> **Five principles to aid healthy and supportive touch in relationships with children**
>
> Babies. Monkeys. Parents. They all provide a slightly different twist on a similar conclusion: Touch is a mighty contributor to our origin stories, with a lasting impact on our biology and behavior. The touch we receive in childhood can impact how we learn, bond, and interact with the world around us. Tactile experiences in early life set the stage for what touch means to us as we grow. They contribute to the consequences of touch for broader behaviors, consequences that matter to our health, well-being, and day-to-day lives.
>
> Still, we cannot deny that touch during early life can be challenging to navigate. As we will see in later chapters, preferences for touch are not universal. This is true for children as much as adults. Unhealthy experiences of touch carry risks throughout life.
>
> Caregivers and those who work closely with children can face the difficult task of ensuring that children have access to supportive touch while balancing the risks of unintentional outcomes of unwelcome touch during childhood. Below are some considerations, adapted from the educational organization Penn State Extension[16] and the nonprofit organization

Zero to Three,[17] that those involved in childcare may wish to consider in aiding healthy and supportive touch when interacting with children.

1. **Reflect**. What we've seen in this chapter is that our early tactile experiences can have a lasting impact. Later, we will see that it is probable that our own childhood experiences impact how we think about touch. To help us engage with children we care for, it can be helpful to reflect on our own thoughts about touch, and how these thoughts may contribute to how we choose to interact with others we care for. Whenever we touch someone else, we strike a balance between our past, present, and future interactions. Being mindful of our own thoughts about touch can help us start thinking about how we engage with others.
2. **Ask**. Asking permission and understanding another person's sensitivities around touch can help us better understand their relationship with touch. This is true whatever the age of the person we interact with. Some schools try to encourage this culture by allowing kids to choose how they greet each other at the start of a school day: a high-five, a fist bump, a handshake, a hug, or not at all. This helps them to learn about their body freedom and consent from an early age.
3. **Learn**. Staying educated about healthy and supportive touch behaviors can help caregivers and people who work in caring professions. Learning about how preferences for touch can vary between individuals helps us consider how we approach touch with others. If you work with children, involving and learning from their families can provide valuable information about their unique requirements.
4. **Go slow**. When we talk about touch, people sometimes assume that more is always better. This is not automatically

the case. Supportive touch is about sharing caring tactile experiences that are correct for the person you are interacting with. This means that you do not necessarily need to be touching all the time. Nor is a rapid increase automatically going to be the right action. Reflect, ask, learn; *gradually* attempt to match healthy and supportive touch in a way appropriate to that person's needs.

5. **Educate.** Once we understand a child's approach to touch, we can feel confident in educating others about those needs. Sometimes a child might not want to immediately hug a relative they've not seen for a while, or they might find hugs overwhelming full stop. It can help to explain this to people. Education also involves teaching our children about differences in touch preferences. We can educate about supportive touch by modelling respect towards each other's body boundaries. For instance, talking to children to let them know before touching them—saying things like "I'm just going to lift you up now" rather than simply touching and lifting them without warning.

CHAPTER 2

Scientists Who Stroke: The Neuroscience of Touch

If you were ever to visit my lab, you could be excused for thinking that I model myself on Doc Emmett Brown from *Back to the Future*. For those who have not seen this eighties classic, Doc Brown is a fictional scientist and the inventor of the DeLorean time machine. I can only wish to emulate his success—including owning a DeLorean. But my lab isn't far from the mix of gadgets you see in his garage workshop.

I own an assortment of equipment that has been accumulated over the years. There are lab coats here and there; mannequins, robots, fake hands, toy worms, and slime; and I'm sure I own every type of brush you might need. Much of this is from a study where we tried to convince people they were touching things like maggots.

So many gizmos and gadgets—I literally have something called a multiple pulse gizmo in front of me right now. However, having said all this, if you were to visit my lab, it's possible that what might first capture your attention is what's written on the key ring as you enter the door: *Scientists who stroke.*

My research team and I make a living by stroking and sometimes hugging other people. We are part of a rapidly expanding group of scientists who want to understand how we judge whether touch feels good or bad, and how those judgements impact our social wellness and health.

We study affective touch—a touch with a pleasurable or emotional component to it. One of the ways we do this is by bringing people into our lab and stroking them. Slow and gentle stroking, in a manner that mimics the gentle caresses you might feel outside the lab from a partner, caregiver, or friend.

The reason for this connects back to those CT fibers we came across in our discussion about how touch develops, the receptors in the skin that are sensitive to gentle stroking.

Affective-touch scientists around the globe have devoted considerable amounts of time to determine how our CT fibers transmit information about affective touch on our skin to the brain. If someone strokes your arm, you may think, "This feels nice." Our job is to ask precisely why. What is the process that leads you to believe this? In layperson's terms, the question we are asking is: What makes touch feel pleasant?

THE SCIENTIST AND THE CAT

To help us understand the answer to why stroking can feel good, let's look at what our CT fibers respond to. You might recall from the previous chapter that CT fibers appear to love a touch that resembles a gentle caress. To understand the detail behind this, we first need to focus on the different types of skin we have on our bodies.

Broadly, there are four types of skin on the human body: skin at the junction of membranes like that found around our lips and tongue; skin that lines the inside of body orifices; skin without hair; and skin with hair. For our purposes, I want to focus on the last two types: our hairy and non-hairy skin.

Look at your hand right now, back and front. You'll probably see that your palm is non-hairy, but the back of your hand is hairier. The palms of our hands are often without hair—more formally known as glabrous skin. This contrasts with most of the rest of our body, which is quite hairy: We should remember that just because hair is not always visible doesn't mean that the skin is not hairy somehow.

Arms, legs, face—you'll likely find hairy skin here. Research suggests that more than 90 percent of the body is covered by hairy skin. And it is in hairy skin that our CT fibers thrive.

Our knowledge about CT fibers began over 80 years ago. They were first documented in cats by Yngve Zotterman from the Karolinska Institute in Sweden.[1] I wish I could give a warm story as to why studies of CT fibers started with cats. Sadly, studying cats was not because of a desire to determine whether they liked to be stroked or not. Instead, it reflects that before the 1980s, they were simply the right size for experimentation in neuroscience, as the equipment was too big to be used on smaller animals.

Decades later, in 1990, Magnus Nordin of Uppsala University Hospital discovered similar gentle-touch-responsive fibers in the hairy skin of humans.[2] The way he did this is unlikely to appeal to the squeamish. Volunteers in the study had their skin pierced to enable microelectrodes to be inserted just above the eyebrow. This allowed Nordin to record from participants' supraorbital nerve—a sensory nerve connected to the forehead and upper eyelid.

Taking part in a study like this might sound a bit gruesome to many of us. Thankfully, because of the brave people who chose to volunteer, Nordin comprehensively demonstrated the role of CT fibers in humans. Slow stroking was particularly effective at activating CT fibers, whereas fast stroking reduced the response. Essentially, we are hardwired to process slow, gentle touch.

Research that has followed Nordin's study has identified optimal speeds to trigger responses from our CT fibers and to evoke feelings of pleasantness in the person being touched: People tend to find stroking most pleasant when performed at a speed of approximately 3 cm per second; this is touch that can be best described as low-force, slow-speed, caress-like touch.

Our CT fibers respond less optimally if the person stroking us slows down or speeds up too much. In that case, we find their touch less pleasant and sometimes even irritating.

In other words, the caresses we like the most map onto the preferences of CT fibers in our hairy skin. This response to stroking speed has been shown from childhood through adolescence to adulthood. Regardless of age, we tend to react similarly to a preferred touch speed.

Findings like this have led many scientists to conclude that positive and pleasant aspects of touch are often conveyed through touch that optimally stimulates CT fibers.[3] The biology of why a gentle caress can feel good starts by targeting nerves in our skin that hunger for this type of touch.

Did you know?

We show preferences for gentle stroking into the ninth decade of life, even when other aspects of tactile sensitivity decline.[4]

THE LITTLE CREATURES IN YOUR BRAIN

There are many ways we can experience the touch of another as pleasant, and the fact that we have fibers in our hairy skin is just one part of this story. There is much more complexity to this process than receptors on the skin. You might be able to think of a situation where a partner stroking your arm felt lovely. However, I suspect that if that partner were doing the same thing when you were five minutes from an important deadline, you might not enjoy it as much.

Various factors determine how you will perceive touch from another person. This could include whom the person touching you is, the context of the interaction, and even your past experiences of touch that day or years before. To answer questions like "Am I enjoying being touched by this person right now?" it's evident that we rely on more than just CT fibers alone.

Before I go into the finer details of how we experience pleasant touch, let me start by addressing what, on the face of it, is a more straightforward question: What happens in the brain when we are touched?

I'd like you to start by picturing a scene: You're on the platform at the train station, waiting to begin your daily commute. It's busy this morning because of travel delays. A fellow commuter needs to get past you. You're looking the other way with your music playing through your headphones. They tap you on the arm to get your attention so they can squeeze past you. How do you know that you've been touched?

The answer to the question might appear simple: "I feel it." Yet behind even the most inconsequential tactile sensation is a host of biology, past experiences, and expectations.

As we've seen already, we experience touch through nerve fibers in our skin, such as our CT fibers. Still, CT fibers aren't the only nerves involved in sharing tactile experiences.

Our skin is thought to be the largest sense organ in the body. It's filled with various nerve fibers and receptors that detect all sorts of tactile sensations: pressure, temperature, vibration, and itch.[5]

One group of receptors is known as our low-threshold mechanoreceptors. While mechanoreceptors may sound like something out of the *Transformers* movie, their function is not to support giant alien robots. Instead, they help to translate external tactile features in the world.

There are four mechanoreceptors of interest for our purposes: Meissner's corpuscles, Merkel's discs, Pacinian corpuscles, and Ruffini endings. These contribute in different ways to feelings of vibration, flutter, pressure, and the sense of our skin stretching.

And a bit like all good bands of four—think the Beatles, Metallica, or Queen—these four low-threshold mechanoreceptors bring different functions to the party to help create a sensory masterpiece that enables us to discern the tactility of the world around us.

When someone prods your shoulder to get your attention, your low-threshold mechanoreceptors will likely detect it. They send rapid signals to a part of the brain called the thalamus: If touch were a USB stick, you could think of the thalamus as the docking station. The thalamus collates the information for our touch receptors and sends it to other parts of the brain.

The somatosensory cortex is one of the critical brain areas to which the thalamus sends information about touch; it's a central part of how the brain processes touch experience. Some might even describe it as the brain's mailroom for receiving tactile signals from different parts of our bodies. A poke on your shoulder? The somatosensory cortex hears about it. A tickle of your feet? The somatosensory cortex knows about that too. A toy fish tapping your bum—okay, enough about that in Chapter 1—but yes, the somatosensory cortex also knows about this.

One of the intriguing things about the makeup of the primary somatosensory cortex is that it has a dedicated representation of each body part that receives touch. A section for touch to the hands, a section for touch to the face, a section for touch to the feet, and so on.[6]

These distinct representations are wrapped around the middle of your head. The easiest way to describe the location is that if you were wearing headphones, the somatosensory cortex is situated roughly where the padding of the headband would be. Your feet are represented close to the middle of your head. Your face is represented closer to your ears.

The tactile body map in the brain differs from how our body is structured in the real world. The brain dedicates more space to areas that typically receive more touch or have more touch sensitivity, like our lips and fingertips.

Scientists measure our touch sensitivity by simultaneously applying touch to two locations on a given body part's skin. They reduce the distance between these two points, bringing them closer and closer together until people can only detect one location being touched. Two touches still occur, but the scientists want to see how close together they must be for people to misperceive those two touches as a single touch. They want to know when people stop detecting the difference between each separate touch.

People are said to have greater tactile sensitivity when they can correctly perceive two distinct touches that are applied very close together on the skin. Our touch sensitivity on the fingertip can be less than five millimeters. In contrast, sensitivity on our shoulders can be more than thirty millimeters apart. These differences reflect the fact that we need to have more refined touch acuity in our fingertips than on our shoulders.[7]

Did you know?

Braille readers show greater representations of their reading fingertips in the somatosensory system. This adaptive quality of the brain reflects the lived experience of Braille readers, who are using their fingertips more than other people. It has also been shown that people born blind (or who lose vision shortly after birth) recruit a more extensive network of brain regions than sighted people when performing tasks that involve perceiving tactile information. Effectively, their brains remodel to account for differences in their sensory world. For example, parts of the brain typically recruited for vision by sighted people can be activated by touch in people who experience blindness in early life.[8]

As noted, the somatosensory cortex gives more space to body parts with greater tactile sensitivity. To make this a bit easier to conceptualize, we can use a real estate analogy. If our somatosensory cortex were a strip of land in our brain, we could say that each body part has a dedicated building on that land. There are many body parts to fit on one strip of land, but limited space. Some parts get more space than others.

The body parts that demand more attention in terms of their importance for engaging and sharing touch get more extensive properties. Our hands might be considered the size of a mansion, while our shoulders are more like a tiny cottage or studio apartment. This means that if our body resembled how it is represented in our brain, it would look very different from reality. We'd have giant hands, small necks, big lips, tiny bellies, large fingers, and so on.

Scientists call this the sensory homunculus. A quick internet search of "sensory homunculus figures" will let you see depictions of the strange way we might look if our bodies matched our brain representations. These surprising and disturbing creatures show us in odd-looking proportions, a bit like our reflections in a funhouse mirror at a fairground.

The somatosensory cortex is vital in making us aware that we are being touched. Yet it is only one part of a dynamic brain network that contributes to feelings of pleasurable touch. Brain imaging studies indicate that pleasant forms of touch—like CT-optimal touch—are associated with changes in brain activity in a part of the brain called the insula.

In many brain diagrams, the insula appears in the shape of a teardrop, buried deep in a fold beneath the frontal, temporal, and parietal lobes. It becomes active when adults and young babies experience pleasant forms of touch, like slow and gentle stroking. For instance, in 2019, researchers from the University of Turku showed that the insula became activated following gentle skin stroking in infants aged 11 to 36 days old, another reminder that we respond to the different types of touch we receive from our first moments on the planet.[9]

The insula contributes to more than just pleasant touch. If you've ever taken the time to focus on how your body feels at a particular moment, you likely have engaged your insula. It has been connected to various functions, including regulating the nervous system, integrating sensory information, making us aware of our emotions, and helping us understand bodily signals such as temperature, pain, or thirst as well as sensations of hunger and fatigue. In short, the insula contributes to many everyday functions connected to our health, well-being, thoughts, and emotions—including our experience of pleasant touch.

In adults, pleasant skin stroking tends to lead to brain activity within the posterior insula, a region towards the back of the insula.[10] It also increases connections between the posterior insula and other brain areas with which the insula communicates. This can include networks of brain areas involved in activities like integrating emotions and regulating and perceiving key bodily signals.

Pleasant slow stroking also activates brain regions involved in perceiving social signals. An example of this is the superior temporal gyrus. This brain region can play a role in making social judgements about other people, such as their intentions or traits. When people find slow stroking particularly pleasant, they are more likely to activate the superior temporal gyrus. In other words, when we perceive pleasant touch, we recruit areas of the brain involved in judging social intentions more.

It's intuitively easy to picture why this might be. Pleasant touch experiences are both sensory and emotional events. Sure, we might enjoy the physical feeling of being stroked, but there will also be emotional, motivational, and social factors that contribute to making pleasant touch feel good. For instance, past experiences may contribute to different emotional reactions to slow and gentle stroking.

This all means that when we experience pleasant touch, we recruit brain regions that are in anatomically favorable positions to influence a range of functions that can aid these types of evaluations.

WHY PLEASANT TOUCH IS LIKE ICE CREAM

Rewards are an important part of life. Just think of how you use them to motivate you to engage in behaviors you might prefer not to. The postworkout treat that gets you to that 7 a.m. gym class on a Saturday might be one example. It turns out that pleasant touch can be rewarding to the brain too.

To understand how and why touch is rewarding, we first need to know how our brain processes rewards. A network of brain regions is involved in processing naturally rewarding features of our environment. I'd like us to focus on two: the orbitofrontal cortex and the striatal area. These brain regions are part of our neuroanatomy involved in processing our experience of natural rewards like food or sex.[11]

To illustrate, think about eating something that you might reward yourself with after a hard week at work—like your favorite ice cream sundae, perhaps. In this situation, your orbitofrontal cortex is likely to be a key brain area in detecting that you're eating something you enjoy. It decodes that the taste of ice cream is present and tells your brain that you're experiencing something rewarding. This doesn't necessarily have to be ice cream: The orbitofrontal cortex has been shown to track other rewards, like erotic stimuli, money, and smell. In short, we can think of our orbitofrontal cortex as a reward tracker.

In addition, it helps us learn actions we might want to engage in to maximize rewards. The orbitofrontal cortex does so through interaction with striatal brain areas, which also receive input from the insula.

Striatal regions are found deep in the brain. They receive signals from hormones associated with movement, memory, pleasurable rewards, and motivation. They also have the orbitofrontal cortex on speed dial. Part of the reason for this is that the orbitofrontal cortex and striatal brain regions share information about rewards and how we might want to approach them to maximize engagement. To connect this back to ice cream, if we identify an ice cream sundae that we find rewarding nearby, we want to know how to maximize the likelihood of gaining that reward.

The rewarding value of ice cream is clear, but what does this all have to do with touch?

In 2016, a team of scientists from the University of Gothenburg showed that experiencing pleasant touch triggers brain activity in the reward-related brain circuits of the orbitofrontal cortex and striatal brain areas. They achieved this by diligently stroking people with a soft brush for 40 minutes inside a brain scanner: a long time to be stroked, I know!

Throughout the stroking experience, the participants rated how pleasant they found it. Experiencing pleasant touch triggered brain responses in the posterior insula, the orbitofrontal cortex, and the striatal brain areas.[12] Like taste or sex, pleasant stroking evoked brain activity in reward-related brain systems.

Why would affective touch be experienced in a rewarding way? The answer to this may be linked to the fact that touch that optimally targets CT fibers has also been related to hormonal releases that can contribute to positive feelings. One example is the release of endogenous oxytocin—a hormone associated with relaxation, calmness, and the neurobiology of close social relationships. Some suggest that pleasant touch may act as a reward by allowing us to de-stress and form bonds with others.[13]

To understand why, it is helpful to be aware that oxytocin can inhibit sympathetic nervous system activity. In contrast, it can support parasympathetic nervous system activity. In simple terms, our sympathetic nervous system prepares the body to respond to stressful situations, whereas the parasympathetic nervous system can calm our body back to baseline. By contributing to oxytocin release, pleasant touch may therefore contribute to supporting the parasympathetic nervous system bringing the body to a state of relaxation and restoration.

Limiting any complex human behavior to one system or hormone is never straightforward. The dynamics of human social interaction are complicated, so understandably the brain basis of pleasant touch goes beyond just oxytocin release. Current thinking suggests that an interplay of different hormones and receptors is important. In addition to oxytocin, other hormones like dopamine and serotonin are thought to play an important role in making affective touch rewarding. Dopamine impacts how we feel pleasure and motivation. Serotonin contributes to feelings of well-being and happiness. Both are associated with other behaviors like sleep, memory, and learning. When we engage in forms of social touch that help build bonds between us—like a caress—it is likely that a mixture of dopamine, oxytocin, and serotonin is released.

THE UNFRIENDLY STRANGER

Amidst all this talk about stroking and CT touch, something might be bugging you. Sure, CT-optimal touch on hairy skin can feel nice. But many other interactions involving touch don't include our hairy skin.

You might enjoy someone tickling the soles of your feet. You might enjoy the feel of a kiss on your lips. Even something as simple as shaking hands may be fun—yes, some people enjoy this feeling. In short, stroking hairy skin is not the only way we can enjoy pleasant touch.

To understand how different forms of touch contribute to tactile experiences, it is helpful to consider a subtle distinction between social touch and CT-optimal touch. For our purposes, we can think of social touch as a broad umbrella term that covers all sorts of touch between people: It includes CT-optimal touch like stroking, but it extends to other types of touch too. Holding hands, a pat on the back, or a supportive hand on the shoulder are all examples of social touch. All can influence social interaction and, as we'll come to see in later chapters, social wellness.

Social touch is influenced by all the nuances of social interaction that fill our daily lives. How we respond to touch from another person can change in an instant. You might imagine that our brain responses will likely alter based on contextual factors too, and you'd be right.

We will come to many of these contextual factors in Part 3 of this book. For now, an important aspect for us to focus on is *who* is touching us.

Brain responses in our somatosensory cortex can change depending on whether we perceive the person touching us to be happy, angry, or sad. Their smell can change how we perceive the pleasantness of their touch, leading to differences in how the insula responds to affective touch.

Studies on babies have also shown that reductions in heart rate connected to social touch can change depending on whether it is their caregiver or a stranger who is stroking them. In one study, nine-month-old babies were stroked by either a parent or a stranger. The infants' heart rates only decreased when stroked by their parents.[14] Even in the first year of life, our relationship with the person touching us counts.

You may think this experience of relaxing touch is just something to do with parent–child interactions. Yet adults show a similar pattern. Do you recall that parents tend to spontaneously stroke their children at optimal speeds that target CT fibers? When asked to stroke partners or friends, similar results have been found in adults. Curiously, when we are asked to interact with strangers, this speed changes—we speed up our stroking.

It's too early in our understanding of these behaviors to suggest a strong reason why this might happen. We might expect that it has something to do with society's unwritten rules about interacting with people we do not know well. We may just feel more awkward stroking someone we don't know. Whatever the precise mechanism, there is a good chance that we will be less likely to spontaneously provide CT-optimal touch to strangers than to our family and friends. Our social world is full of nuance. And so it is only natural that our brain response to touch is nuanced too.[15]

DON'T TOUCH YOUR FACE!

For our brain to know that we are engaging in social touch with another person, it needs to distinguish between a touch that comes from another rather than from ourselves. Self-touch is an incredibly prevalent form of touch that we engage in daily. Out of interest, how many times do you think you've touched your face in the last hour?

A comprehensive review conducted by epidemiologists at the University of Auckland suggested that, on average, self-face-touching occurs over 50 times an hour.[16] What is more, touching the facial T-zone (eyes, nose, mouth, chin) happens even more often—over 65 times an hour based on averaged data. If anyone reading this estimated a similar number, you did far better than me—and I claim to have a good awareness of touch!

We touch ourselves for a whole host of reasons. Sometimes it can be spontaneous, like flicking our hair with our hands when we are holding a conversation with someone. Other times it can be more deliberate, like if we bump part of our body and try to rub it better.

Our brain responds very differently to self-touch than to receiving touch from other people. One famous example of this is attempting to tickle yourself. If you have someone close to you while you're reading this, you could give it a try.

First, try to tickle your feet yourself. Now ask your friend, partner, or family member to do the same. Many people will find that the sensation feels different—more pleasurable (or agonizing, depending on your preference) when someone else tickles you than when you tickle yourself.

Whether we can or can't tickle ourselves is not just a question for home science projects. Scientists have shown that our somatosensory cortex responds differently when we touch ourselves as opposed to when we are touched by someone else. Our brain activity around touch is dampened down by self-touch.

A suspected reason for this is that other parts of our brain predict the outcome of our next move more accurately than when someone else does the touching.[17] This means we are less likely to feel the same sensations from self-touch because our brain has a better idea of what is coming next.

Incidentally, scientists suggest that there are two types of tickling from others. In one type, light tickling can sometimes make people feel like they want to scratch their skin. In another, deep pressure tickling can make them laugh. Enjoyable tickling that makes people laugh can be a stress reliever for some.[18]

The fact that we struggle to tickle ourselves means that we rely on having someone else around to support this process—or the right technology to help us out. In 2022, researchers from the University of Auckland introduced the world to TickleFoot,[19] a piece of technology that uses an actuator and magnetic brushes to tickle the soles of people's feet. So much so that it can evoke laughter.

Photographs of this technology look like someone standing barefoot on a bathroom scale, except the scale tickles your feet. It also comes as an insole for footwear—I guess for those who want to have their feet tickled while wearing slippers? Irrespective, it's an intriguing piece of technology for people who love foot tickles from other people but can't get the same feeling from tickling themselves.

Of course, self-tickling does not account for all the multiple times people touch themselves in an hour. Detailed research has shown some situations where self-touching is more common. If we are stressed or nervous, we tend to self-touch more. This can also be the case as tasks get harder or when we have heightened feelings of emotion. I don't know about you, but I tend to fidget and rub my fingers together more when I'm feeling anxious. Sometimes I don't even realize I'm stressed until I notice my self-touching behavior.

It turns out that my fidgeting quirk may be grounded in science. Some research has suggested that fidgeting can have the potential to act as a behavioral coping mechanism for stress. This work has focused on displacement behaviors—forms of self-contact like biting, pulling, or scratching the hair and skin.

One study from researchers based at the University of Roehampton investigated the contribution of displacement behaviors to stress in men during a potentially stressful experience: performing mental math out loud in front of strangers. This may not sound highly stressful for some, but if you share my high math and social anxiety, you will likely agree that this does not sound like a fun experience at all.

The researchers found that men who engaged in more self-touch during the task experienced less stress.[20] In effect, self-touch appeared to help them do their calculations aloud.

It's not just fidgeting that can reduce stress; some of the latest research suggests that hugging ourselves can help. A study by researchers from Goethe University in 2021 found that a 20-second hug from a researcher or giving a hug to oneself led to reduced levels of cortisol (a hormone connected to stress) compared to no hugs or engaging in non-tactile self-related tasks (more on this in Chapter 3).

*

Researchers have also tried to use self-touch to develop interventions to make us more aware of our experiences and sensations. These researchers study interoceptive awareness: our ability to perceive and understand our body's internal state.

If you've ever stopped to focus on how your heart is beating or to observe your breathing rate, you've engaged in interoceptive awareness. This seemingly mundane task of correctly perceiving our bodily signals is a surprisingly influential contributor to mental health. Better self-awareness of our body's reactions can make us more mentally resilient. And on the flip side, interoception problems, where a person struggles to notice their body's signals, have been connected to mental health outcomes like low mood, eating disorders, and addiction.[21]

Differences in how well people perceive interoceptive signals have been related to how we process emotions and build social connections. And as we know already, these social connections are important to our lifelong happiness and health.

Therefore, finding ways to help improve interoceptive awareness is a high priority. Early-stage findings show that interventions involving regular self-touch or self-body-brushing can aid aspects of interoceptive understanding like emotional awareness and self-regulation.[22] In other words, self-touch can help us feel calmer, more present, and less stressed. These promising results show a fascinating potential for self-touch to be used in interventions to support wellness and self-awareness in the future.

From discriminative touch to affective touch to human connection

Part 1 of this book has focused on the origins of touch. We've seen how touch can contribute to how we form impressions of ourselves and the world around us from our earliest moments. Touch helps us and other animals form bonds that have an ongoing impact throughout our lives. Our whistle-stop tour of different types of touch has shown just how powerful the biological underpinnings of touch can be. Affective, CT-optimal, discriminative, social, and even self-touch critically contribute to our lives.

The distinct biological basis of each type of touch shows how various forms of touch can profoundly impact our brain and behavior.

It's also clear that touch is more than just a matter of nerve fibers. Tactile exchanges are full of complexities. Even in the brain, context matters, and our relationships with the people touching us matter too. The briefest touch can be loaded with meaning and require careful navigation. Our perception of touch can change based on the subtlest cue in our environment. Something as simple as a signal from another one of our senses can change how touch can feel, confusing the brain's ability to receive the signals from the nerve fibers.

What we experience through touch is not always just about touch itself. In the chapters ahead, we'll see how our touch experiences are layered and complex. We will learn how our biological systems respond to this complexity. Our touch biology is a flexible platform for understanding the information we share through our skin. It is a platform for us to build connections with the world and the people around us.

To conclude this section, here are five key facts and implications to keep in mind about the biology of touch.

1. Our skin is the largest sense organ in the body. It acts as a bridge between our environment and the body's internal systems. Some examples of the skin in action include: fighting off allergens, regulating our temperature, protecting against UV radiation, and helping in the production of vitamin D, which can support our health. Taking care of our skin is important for more than just appearance.
2. We have dedicated nerves in our skin that respond to tactile features like pressure, vibration, temperature, and texture. These contribute to the sensory richness of our tactile world. They let us share experiences with others, such as admiring the texture of a friend's clothing or using the feel of items we have collected to share stories about life experiences.
3. We have a particular type of nerve that preferentially responds to slow, gentle touch—like a caress or gentle stroking of the skin. These preferences map onto benefits, with several studies showing that comforting touch can positively affect humans throughout life. So much so that health organizations worldwide have begun adapting approaches to facilitate opportunities for comforting touch.
4. When we are touched, we activate a part of our brain called the somatosensory cortex. This contains distinct representations for each part of our body. It has bigger representations of body parts that are more sensitive to touch, such as our hands and lips. This shows us that the brain adapts to match the tactile world around us (for instance, using tools in our hands can require more sensitivity).

5. Another critical brain region connected to how we share touch is the posterior insula, which contributes to the perception of pleasant touch. Pleasant touch is also associated with activity in biological systems involved in releasing the hormones dopamine, oxytocin, and serotonin. These three hormones are related to psychological behaviors, including bonding, calmness, motivation, pleasure, and reward. As we'll begin to explore in the following chapters, one implication of this is that we might be able to consider touch as a possible vehicle to support some of these behaviors in daily life.

PART 2

GOOD TOUCHING

CHAPTER 3

Healthy Touch: Can Hugs Conquer the Common Cold?

In 2015, I found myself walking through a park in Sydney, jet-lagged, having arrived on a 24-hour flight the day before. I was taking in the views. The sounds of woodpeckers calling in the trees. Yoga mats on the lawn. The sun shining against the beach's backdrop, reaching out to the South Pacific Ocean.

For all the beauty surrounding me, I was deep in thought. A few months before, I had ended a long-term relationship. I'll skip the details; let's just say it had been a difficult time, with a mixture of emotions that seemed to flow through me daily. At that moment, however, there was one clear feeling—for the first time in a long time, I felt alone. Traveling solo. In a place where I didn't know anyone at all.

It was in that moment of isolation that I saw something out of the corner of my eye. A stocky Australian man with wavy greying shoulder-length hair stood opposite me, holding a sign that simply said: *Free Hugs*.

He was part of the Free Hugs campaign, initially launched by another Australian known by the pseudonym "Juan Mann." The campaign had begun in Sydney much earlier than my visit. In 2004, Juan Mann had been feeling depressed and lonely. A hug from a stranger changed everything. "I went to a party one night, and a completely random person came up to me and gave me a hug. I felt like a king! It was the greatest thing that ever happened," he said.[1]

Mann felt it was essential to pay this gift of a hug forward. In June 2004, he appeared in the Pitt Street Mall in central Sydney holding a sign bearing the exact words as the one I'd just seen in the park: *Free Hugs*. He was giving strangers who needed it a bit of a boost to their day through a hugging embrace.

The story of the Free Hugs campaign is fascinating. What began with Juan Mann moved forward to include more and more individuals who offered hugs to strangers in public places across the globe. The movement progressed from initial distrust to engagement, involving legal battles, petitions, a rock band, and even Oprah![2]

On that day in 2015, though, none of this history mattered to me. At that moment, when I was standing in front of a stranger holding a sign offering a free hug, all that went through my mind was: Why the hell not? I had the hug. Like Juan Mann, the simple act of kindness changed my mindset and helped me on my way.

THE HUG THAT BEAT THE FLU

Hugging is a very powerful tradition in many cultures. Research has shown that young adult couples report that they cuddle four or five nights per week before sleep. This simple act may be relatable to many people. After all, most adults tend to sleep in the same bed as their romantic partners, and spooning is one of the most common positions in which people report falling asleep.

Outside the bedroom, many people regularly hug family and friends throughout everyday life. We use hugs as a greeting. They are a way to show positive intentions upon encountering another person. A 2022 study by researchers from the Netherlands Institute for Neuroscience found that, on average, people hug around six times a day, with the weekends being the days that people hug the most (roughly nine hugs on Saturday and ten on Sunday). The fewest hugs happened on Monday through Wednesday (about four hugs per day).[3]

As you delve into the science of hugging, you'll find all sorts of hugs described: the bear hug, the from-the-back hug, the polite hug, and even the unfortunate one-sided hug, where the other person doesn't hug you back.

Preferences toward hugs can vary. When people are asked to rate videos of the intimacy of hugs between other people, researchers found that different data patterns emerge according to the type of hug, the hug duration, and the gender of the huggers.[4]

One study found that for two men, the crisscross hug is viewed as the most intimate when it takes place for one or five seconds but the least intimate when it takes place for three seconds. A crisscross hug involves people crossing their arms over each other's shoulders and waist. One arm goes up, and one arm goes down, creating an X with our arms.

In contrast, the engulfing hug is considered the most intimate for two women when the hug is as brief as a second. In an engulfing hug, one person folds their arms inwards while the other wraps their arms entirely around them. You might find someone engaging in this type of hug when consoling a crying friend.

This is only part of the story, however. Our perceptions of the intimacy of hugs between women change as the hug duration continues. For slightly longer hugs, such as three-second hugs, crisscross hugs are viewed as the most intimate. While at five seconds, neck–waist hugs become most intimate. In neck–waist hugs, one person hugs the shoulders of the other person while the hugging partner wraps their arms around the waist of their fellow hugger.

Hugs also play an important role in expressing emotions. I'm sure many of us can think of a range of situations where hugs have helped us communicate with other people, perhaps as part of a celebration, in sympathy, or expressing gratitude. Hugs can be an emotionally meaningful touch for those who share them.

The man who hugged me in Sydney perhaps knew this better than I did. After all, emotionally meaningful hugs are core to the origins of the Free Hugs campaign.

Now I know that not everyone will find hugging a stranger in public appealing. I can see how some people may describe Free Hugs as a bit of a fad. But perhaps we can at least agree that even if you don't choose to hug a stranger in a park, many of us are instinctively drawn to hugging as a way to comfort and communicate.

Think about it for a moment or two: Can you think of a hug that you remember because of its positive emotional effect on you? What was special about it? What made it a good hug?

Seventy percent of people who completed the Touch Test reported that they could remember a hug because of its positive emotional effect. The most common reasons for this were the nurturing and supportive aspects of the hug. As one person explained, "Whenever my mum hugged me, I would feel safe, warm, cared for, understood, loved, praised . . . whole."

For many people, hugs were reported as a highlight of life. This was captured in comments like "One of my favorite things is people who are good huggers or whom you hug well with—it is a very special thing. I can count the people on one hand who are great huggers. When a hug is special, I can tell you exactly where I was and who gave those hugs to me."

There are good reasons for us to remember hugs for their positive emotional effect. One study of American university students showed that engaging in at least five daily hugs over four weeks increased self-reported well-being.[5] In this research, the well-being of one group that engaged in at least five hugs per day was compared to a group that was asked to read each day. The hugging group showed positive changes in their self-reported well-being. The reading group did not.

As an aside, my editor and I are naturally appalled at the prospect that reading did not help with well-being in this instance. However, in other studies, reading has been related to benefits for health and well-being. The reading material type can make a difference, though: One study found that reading can provide an advantage for longevity in older adults, but that book reading was more advantageous for survival than newspapers/magazines.[6] Well done for deciding to learn about hugs *and* read a book—a double whammy of benefits.

Other research has found that hugging can help reduce adverse experiences in daily life, such as arguing. Researchers based at Carnegie Mellon University interviewed people from the Pittsburgh area for 14 consecutive days about whether they had experienced any conflicts with other people on the days in question. They were then asked about how positive or negative their moods were. They were also asked whether they had hugged anyone each day.

After more than 400 painstakingly conducted interviews, the responses for over 5,000 days of observations about hugging, mood and conflict were analyzed. As you might imagine, experiencing conflict with someone during the day was linked to more negative feelings in the study participants. An argument left people feeling crummy.

Importantly, however, it was seen that experiencing a hug on the day of the conflict helped buffer against the negative emotions. The research showed that receiving a hug benefited the participants' moods on days when arguments took place.[7]

Miles away from Pittsburgh, scientists from the University of North Carolina examined how the frequency of hugs between partners was linked to physiological stress markers in women. They found that more frequent partner hugs were associated with higher circulating oxytocin levels and lower blood pressure.

You might recall from the opening chapters that oxytocin is a hormone associated with relaxation, calmness, and the neurobiology of close social relationships. Using detailed data analysis techniques that enabled relationships between different variables to be determined, the team of researchers in this study showed that more frequent hugging was linked to differences in physiological measures of stress—a relationship partly due to the interplay between hugging and circulating oxytocin. In short, more hugs, less stress.[8]

Beyond stress, hugging has been linked to other physical health benefits. One notable study was published in 2015 by a team of researchers based at three major American universities.[9] This work, led by Sheldon Cohen of Carnegie Mellon University, involved the same group of people from the Pittsburgh area that we came across in the experiment showing that hugging can help reduce the adverse effects of interpersonal conflict. This time, the team was less interested in what happened after an argument. Instead, they wanted to know how hugging contributed to developing symptoms of a virus.

The researchers recorded whether people received hugs or not for the 14 days of the study. Subsequently, the participants were quarantined and exposed to a cold virus.

The development of infection symptoms was measured. This process involved days of measuring nasal secretions, daily mucus levels, and nasal clearance times to assess evidence of viral infection (apologies if you're reading this while eating). The researchers wanted to check how the amount of hugging an individual received before quarantine might impact the development of these illness symptoms.

What they found was pretty amazing and somewhat dramatic. The amount of hugging that took place before exposure to the virus protected against the risk of developing some illness symptoms. Higher rates of daily hugging predicted more efficient nasal clearance in the study participants.* In this study, hugged people were healthier people.

It's important to recognize that the reasons for this may not have been simply about the sensory aspects of the hug. Sure, hugs can feel good, but the critical factor may be more about what hugging means to us.

Case in point: In the above study, having more perceived social support (measured with a questionnaire) also protected against the rise in infection. Using comprehensive data modeling, the authors showed that the buffering effect of hugs could explain some of the beneficial effects of perceived social support. In other words, hugging may help fight viruses because it implies social support to the person receiving the hug.

As the authors of the research state: "The apparent protective effect of hugs may be attributable to the physical contact itself or to hugging being a behavioral indicator of support and intimacy."[10] Either way, snuggling up in the cold and flu season might not be such a bad idea.

Feeling increasingly vindicated in my decision to take a free hug, I stumbled across an article in 2021 that had me firmly convinced that consensual hugging was good. Researchers from Sam Houston State University and the University of Arizona published a study showing that daily hugging was predictive of lower markers of chronic inflammation.[11]

For those who do not know, inflammation is a natural response that is critical for survival. It can produce symptoms like fever or swelling associated with attacking infections.

In many situations, acute inflammation can be adaptive for health. For instance, if we cut our finger, it swells up as our

---

* It should be noted that not all measures changed in response to hugging. For example, rates of hugging were unrelated to mucus production.

immune system kicks in. Unfortunately, our inflammatory response can sometimes become chronic: Our body might stay on high alert when it doesn't need to. This can contribute to various diseases considered among the leading causes of disability and death worldwide. Cardiovascular disease, cancer, diabetes mellitus, and autoimmune and neurodegenerative disorders are just some diseases connected to chronic inflammation. Understanding ways to lower chronic inflammation is therefore important.

The researchers, Lisa van Raalte and Kory Floyd, wanted to tackle this problem by studying hugging. They wanted to know whether the amount of hugging people had in their lives impacted chronic inflammation. Like the work of Sheldon Cohen and his colleagues, the hypothesis here was that greater social connection facilitated through hugging might lead to better health outcomes.

To tackle this prediction, van Raalte and Floyd asked undergraduate students about their hugging behavior daily for two weeks. They also took measures of inflammation from each participant. They did this using saliva samples to calculate pro-inflammatory cytokine levels (this is one way of measuring chronic inflammation, with higher values related to elevated inflammation).

The results were astonishing. More frequent daily hugging was related to lower pro-inflammatory cytokine levels in the study participants. In other words, frequent hugging was linked to lower markers of chronic inflammation and might hold promise for better health. More work is needed to extend the results beyond one group of undergraduate students, but it could have wide-reaching implications as a low-cost way to help millions of people.

The results also align with a broader message from studies across different decades: The simple act of hugging can have surprising impacts on markers of health and well-being. This is seen in self-reported wellness measures and physiological characteristics of stress responses, relaxation, and immune system functioning. It seems that hugs do more than connect us in the moment of an embrace; they may carry positive benefits for our health and well-being long afterwards, too.

THE TOUCH OF YOUR HAND

Before moving on, it's essential to pause and think about the implications of what we have just read. Caring touch, like hugging, might be able to trigger physiological responses that are helpful for our health and well-being. That's a bold statement. It's incredible to think that something as simple as a daily hug—an everyday greeting between partners, family, and friends—can have such a powerful effect.

As compelling as the evidence is, there is a significant caveat to the studies mentioned so far. They all tend to involve healthy adults based in America, and often university students. That's not necessarily a problem, but it is also not fully representative of the diverse nature of global society.

It speaks to a broader concern that we must be aware of: Much psychological science research is full of WEIRD findings. When I talk about WEIRD results, I do not literally mean weird as in strange and unusual. The acronym refers to people from white, educated, industrialized, rich, and democratic backgrounds. WEIRD people tend to be recruited to research more than other groups. One famous study in 2010 reviewed databases of comparative social and behavioral science studies and found that WEIRD samples represent as much as 80 percent of study participants in research but only 12 percent of the world's population.[12]

In other words, people from WEIRD backgrounds are overrepresented in some research samples. A risk of this overrepresentation is that our understanding of human behavior becomes skewed towards what happens in WEIRD people. It makes it challenging to generalize research findings to other groups that do not match these background characteristics.

This is a problem for scientific topics in general. It is particularly acute for research about touch. As we'll come to see repeatedly throughout this book, our experience of touch is nuanced and can vary from person to person. While a hug may be comforting for some people, it can be intrusive and stress-inducing for others.

This is true not just for hugging but for many different forms of touch. Even something as seemingly innocuous as a supportive hand on someone's arm can mean or feel very different depending on our past experiences and history of touch.

An important caveat to all the evidence we discuss is that none of it should be taken as evidence that touch is equally beneficial to everyone. Attitudes and experiences can fluctuate between us. This means that the outcomes of touch are also likely to change from person to person and from one context to another. We'll return to this in Part 3 when we talk about the nuanced nature of touch. For now, let's keep these nuances in mind, especially as we start to look beyond hugging to consider how other forms of touch might impact our health and well-being.

So, what about those other forms of tactile interactions? Sure, hugging appears to have positive outcomes, but does holding hands have a similar effect? How about a massage? Or simply having a friend put their hand on your back when you meet them in a coffee shop? Does it matter who does the touching—a friend, a nurse, or a stranger?

It turns out that extensive evidence shows that a variety of forms of welcome and appropriate touch between people can be linked to benefits for health and well-being in healthy adults. This is particularly true in response to stressful situations. Various types of touch—such as massage, kissing, and even hand-holding by a partner—have been shown to benefit people in such situations.

For instance, some research indicates that getting couples to increase the frequency of romantic kissing in their relationships can improve perceived stress, relationship satisfaction, and even cholesterol levels (more on this in Chapter 5).

Other studies have indicated that touch may be more effective in reducing stress than verbal support.[13] In this work, led by researchers from the University of Zurich, young adult women were exposed to a stressful situation—public speaking and completing mental math out loud in front of people. Fifteen minutes before they did this, they received either a neck and shoulder massage from their partner or verbal support alone. One group of women also engaged in no interaction with their partners. The results demonstrated that touch intervention reduced stress hormone levels and heart rate response in the young women whose partners gave them a massage before the stressful situation. Verbal social support alone did not. Put simply, a helping hand made a difference. The next time I'm giving a talk at a conference or a festival, I'll be sure to bring my partner along for a neck and shoulder massage beforehand.

That hand-holding can help with stress may not surprise you. It turns out that people are pretty good at predicting that touch is an effective way to show support to one another. In 2021, Brett Jakubiak from Syracuse University showed that when people are asked to imagine supporting their partners in stressful situations, they anticipate that touch will be more effective at lowering stress than verbal or no support.[14] In short, people tend to believe touch will be more helpful than other social signals in helping their partners.

Importantly, it's not just touch between romantic partners that can make a difference to health and well-being. In one experiment conducted by researchers from the University of Virginia, brain responses to threatening situations were compared when women held a stranger's hand, their partner's hand, or did not hold hands at all. The threat was an electric shock delivered to their ankle while they had their brain activity measured in a magnetic resonance imaging brain scanner. The women in the study were alerted to the upcoming pain by a cue that told them that the shock would be delivered.

Imagine, sitting in a brain scanner waiting patiently to be told, "We will zap your ankle now." Not my idea of fun!

Thankfully for the researchers and us readers, some people did take part. This allowed the researchers to compare what happens in the brain during threatening scenarios when hand-holding either did or did not happen. It also meant they could compare how the brain responses differed based on the participant's familiarity with the person holding their hand—a partner or a stranger.

The results showed that both partner and stranger hand-holding changed brain responses to the threat of an electric shock. There was less threat-related brain activity when the participant held hands with someone—anyone. Put more simply, the brain activity indicated that participants were comforted by hand-holding compared with not holding hands. It didn't matter if it was a partner or a stranger—both helped. Just think, the next time you reach out for human touch in a scary movie, this is likely not just a random act; it could help the brain diminish the impact of the perceived threat. In a frightening situation, it seems that our response to reach for human touch, even from a stranger, could help our brain respond to the situation itself.

There were some complexities to these findings, though. The study showed that partner hand-holding was more impactful than stranger hand-holding. In fact, the effect of relationship type went a step further: Higher relationship quality with the partner predicted less threat-related brain activity. In other words, the stronger the relationship, the more people gain from holding hands with their partner in a threatening situation.[15]

This reminds me of what we found out earlier in the study by Sheldon Cohen on hugging and health—it is not simply the sensory aspects of touch that make a difference; the meaning of touch matters. You will recall that Cohen's team found that hugging contributed to perceived social support, helping people fight off viruses. It would make sense that when we feel closer to other people, we might feel more social support from them. This may be part of the reason why the strength of the benefits of holding hands can vary based on the strength of the relationship we have with the person holding our hand.

With that being said, we should not lose sight of the fact that touch from a stranger still helped to ease threatening experiences. This carries a variety of implications, like how touch could be used between patients and medical workers in situations where patients are anxious.

It turns out I was not the first to think of this.

In 2001, researchers from the Catholic University of Korea examined the effectiveness of hand-holding in relieving anxiety in patients undergoing surgery.[16] This work involved studying patients who were due to undergo cataract surgery. The surgery took place within the same time window each day. The same surgeon, local anaesthetic, and surgical procedures were used to minimize the impact on the patients of any factors other than the hand-holding they were testing.

Before the surgery, self-reported and physiological markers of anxiety were measured in all patients. What differed was the intervention group that participants were assigned to. One group involved hand-holding with the researcher during the procedure, while the control group did not experience hand-holding. The hand-holding intervention took place during the operation and lasted about 15 minutes. Just before the end of the operation, the self-reported and physiological markers of patients' anxiety were measured again.

Holding hands with the researcher during the operation resulted in less self-reported anxiety. It also reduced levels of epinephrine compared to not holding hands. Epinephrine is a hormone used during the body's stress response; thus, it can be used as a physiological marker of anxiety. Typically, the more epinephrine present, the more anxiety the patient feels. Hand-holding with a professional—the researcher in the study—reduced anxiety in a surgical environment.

I was struck by the implications of these findings. Could hospitals help patients in stressful situations by providing them with a supportive touch from staff? I explored the literature a bit more. Interestingly, the data from the study on patients undergoing surgery in South Korea is consistent with other data available. Scattered through nursing journals from the 1980s to the present day, various reports link touch in treatment settings with reduced state anxiety and stress.*

In 1984, Janet Quinn from the College of Nursing, University of North Carolina, reported that therapeutic touch could reduce state anxiety of hospitalized adult cardiovascular patients.[17] Similar findings of reduced stress and lower anxiety with touch in children have been found in hospital and dental treatment settings in research published in the 1990s.

Even different species have been shown to benefit from touch: Did you know that touch can be beneficial for reducing fear, stress, and aggression in fish? In one study, scientists artificially touched a type of coral reef fish, the surgeonfish (think of Dory from *Finding Nemo*), by using cleaner fish that interacted with the surgeonfish in a tank. For those who do not know, cleaner fish are a type of fish that remove dead skin and parasites from other fish. The scientists

---

* State anxiety refers to anxiety directly related to a specific adverse situation—like undergoing surgery.

wanted to understand how touch from the cleaner fish impacted the stress levels of the surgeonfish.

There's a difficulty with this, however: Asking cleaner fish to manipulate how often they touch a surgeonfish is impossible. So instead, the scientists used mechanical cleaner fish: fake fish explicitly made for the study. In short, the scientists controlled whether the surgeonfish were touched or not.

To determine the surgeonfish's stress levels, they measured their cortisol levels. As we've mentioned, cortisol is often viewed as a stress hormone. The scientists predicted that if touch is important to stress levels in fish, the surgeonfish should show less cortisol when touched by the mechanical cleaner fish.

This was found to be the case. The surgeonfish touched by cleaner fish (even fake cleaner fish!) showed lower cortisol levels than those that were not touched, indicating less stress in touched fish.[18] As the authors conclude, "Our results show that physical contact alone, without a social aspect, is enough to produce fitness-enhancing benefits [for surgeonfish], a situation so far only demonstrated in humans."[19]

All these studies with different people and even different species offer various takes on a similar message: There is a potential for touch to reduce stress and anxiety that is directly related to adverse situations.

You might find yourself wondering how the medical profession has used the information. Do surgeons now always hold hands with patients undergoing surgery, or do dental nurses hold the hands of nervous children? To my knowledge, the answer is no. There is a surprising lack of research on policy intervention in this area.

This is a gap that other researchers and healthcare practitioners have highlighted. A scientific review paper on the neurobiology of touch and stress published in 2022 concluded that "despite the importance of sensory systems in both inducing and inhibiting stress, their therapeutic potential has been left largely untapped."[20] Another review in 2007 sought to assess the effectiveness of therapeutic touch on anxiety disorders. It concluded that there is a "need for randomized controlled trials to evaluate the effectiveness of therapeutic touch in reducing anxiety symptoms in people diagnosed with anxiety disorders."[21]

In a world where people and organizations are regularly looking for low-cost and natural interventions to support healthcare, it would seem touch may have a lot to bring to the party.

THE HIDDEN BENEFITS OF MASSAGE

A final form of touch that has been extensively researched in the context of relationships between touch, health, and well-being is massage. Massage has long been considered a natural method of healing. Some estimates suggest its traditions go back to 3000 BCE, representing millennia of ancient wisdom passed down through generations.

These days, many people are familiar with massage in physical therapy. Like me, you may also seek a massage as a more holistic approach to self-care, relaxation, and restoration.

But massage isn't just an indulgence for the pampered few, or at least it shouldn't be. It has been linked to health benefits from the earliest to the latest stages of life, in healthy people and the sick. The ancient practice of massage has proved to be a powerful way to unlock the potential of touch in the modern day.

One of the world's leading authorities on the impact of massage on health and well-being is Professor Tiffany Field of the University of Miami. The work of Field and her colleagues was seminal in showing the benefits of massage across a range of age and clinical groups.[22]

As we saw in Chapter 1, it has been shown that massaging premature infants in neonatal intensive care units can be linked to weight gain and shorter hospital stays. Just 15 minutes of moderate-pressure massage per day for five days can lead to positive health outcomes for preterm infants compared with those who received standard care without massage therapy.

Preterm infants who received moderate- versus light-pressure massage have also been shown to experience less crying, stress behavior, and fussing. The implication for those practicing massage in preterm infants is that providing moderate-pressure massage can contribute to better outcomes for the infants. This is an approach promoted by some organizations in guidance on how to massage a baby.

We often think about massage from the receiver's point of view, but it turns out that the benefits of massage go both ways. In 2020, researchers from Northumbria University published an article examining how giving and receiving massages in romantic relationships impacted the well-being of stressed but otherwise healthy couples.[23]

Couple massage was beneficial for emotional stress levels and mental clarity, measured via a questionnaire provided to the couples every week. Importantly, it helped both parties in the relationship. It did not matter whether they gave the massage or received it. In other words, simply sharing a massage helped their feelings of emotional stress and levels of mental clarity.

This shared benefit of massage is seen in other relationships as well.[24] It is often reported that massage practitioners experience reduced anxiety after giving massages. Home massage therapists have been reported to show increased confidence, satisfaction, and a greater sense of closeness after treating others. And it has been found that for some retired people, giving a massage to babies can be beneficial for their anxiety, mood, and stress, in some cases more so than receiving a massage from a therapist themselves.

In short, massage can help both parties involved in the experience. It does not matter if you are the giver or the receiver—simply sharing a massage can benefit wellness.

This makes me think back to the No Baby Unhugged initiative that we read about in Chapter 1. We saw there how this type of scheme might help babies get the touch they require early in life. It's a powerful low-cost intervention that could have many beneficial effects on babies in hospitals. I wonder if it might also benefit the volunteers participating in the touching. As we will see in the next chapter, many people report that they need more touch and affection in their lives; they are touch hungry. We know that touch is also connected to loneliness and that being touched by a stranger reduces perceptions of loneliness. I don't know if engaging in programs like No Baby Unhugged helps the volunteers too, but data on the mutual benefits of massage make me wonder if it is possible.

As you dig deeper into the science of massage, you can increasingly find evidence of the benefits of massage on a range of self-reported and physiological wellness measures. For instance, the stress hormone cortisol has been shown to decrease following massage therapy, as has heart rate and blood pressure. In contrast, the neurotransmitter serotonin—associated with helping to support feelings of well-being and happiness—has been shown to increase.

The benefits of massage were really brought home to me in a conversation I had with a lymphedema specialist nurse named Olivia. Lymphedema is a long-term condition that causes swelling in the body's tissues due to an accumulation of fluid.[25] It is common after cancer treatment, particularly after removing lymph nodes. This can lead to distress and discomfort.

One type of treatment for lymphedema is manual lymphatic drainage, a specialized, gentle type of medical massage. Olivia is a trained manual lymphatic drainage therapist. In our conversation, she described to me just how powerful touch is within her profession: "Working with cancer patients showed me how important touch was. For many of my patients, people shy away from them—they either don't want or don't know how to touch them. So, the patients crave touch, but they don't get touch in their lives. They often come to me, having been through a long line of treatments. They will have been through so much, but they won't have been touched. Sometimes, some of my more elderly clients may not have been touched in days or even weeks. Nobody has had time to sit with them and touch them. Our sessions are the first touch that they have had in weeks."

She went on, "By touching them, you can just see them relax. Someone giving them a gentle massage gives a whole stress release. Of course, they gain from the specialized treatment, but they also appear to get so much from a simple touch: just from a gentle massage. The slowing down—the gentle and rhythmic massage—just seems to have this amazing effect on people. I've seen people just burst into tears and feel able to offload thoughts and feelings as soon as you start to touch them. It's like a powerful release through touch. You just see a knock-on effect where the small act of touching can help them come out of a previously dark place."

Olivia's powerful words reminded me of research on other types of massage in cancer patients. Researchers from the University of Colorado Denver compared the effect of six half-hour massage sessions performed by licensed massage therapists with six simple-touch sessions over two weeks in over 350 adult cancer sufferers. The simple-touch sessions involved the patients having hands placed on different body parts (arms, calves, hands, neck) with no side-to-side movement.

It was found that while both groups showed benefits from touch for their immediate pain and mood, massage resulted in the largest improvements.[26] Touch helped—especially massage.

The health benefits of massage are measurable and demonstrated in multiple scientific studies, even in some of the very sickest patients.[27] I can see why data like this has contributed to massage being offered as a complementary therapy in cancer care in some settings.

More work is needed to further explore relationships between massage and health, but these results again point to the potential of touch to impact our health. Putting it all together, you might conclude that if touch can help with wellness, then certain forms of massage take us up a notch. Massage appears to have enormous potential as a turbocharged way to promote positive outcomes for health and well-being.

> ### The healing nature of human touch
>
> Over the months I've spent researching and writing this book, I have often found myself thinking about Bobby Fischer. You will find much written about Fischer's career as one of the world's greatest chess players. He was a Chicago-born, Brooklyn-bred virtuoso. After winning the US chess championship at the age of 14, he went on to become world champion in "the match of the century," the first American to win a title that Soviet-born players had held for 35 years.
>
> Yet it is not my love of chess that keeps drawing me to Fischer. Instead, it is his words, supposedly among his last: "Nothing is as healing as human touch." Looking back over the findings summarized in this chapter, he may not have been that far off the mark.
>
> When you think about it, caring for other people almost always involves touching them somehow. A parent helping a child to get dressed. A child supporting an elderly loved one as they move from one room to another. A nurse touching a patient during treatment. Touch is central to how we nurture and care for each other.

Caring touch comes in many forms. It could be hugging, stroking, or massage. Even simply laying a hand on another person to convey comforting and intimate emotions. These forms of touch have communication in mind. Touch can be a tool for sharing social messages and expressing feelings that show we care. A signal of social support that, as science demonstrates, can help our health and well-being in a variety of ways.

Below are a few take-home messages to help you understand how supportive social touch might be helpful for yourself and those around you.

Hugs can help reduce stress and have positive effects on physical health in a variety of ways. The reasons for this are connected to hugging being indicative of feelings of social support. Getting hugs from others but also giving them yourself may offer benefits for some people.

Hand-holding can also reduce stress and lower pain responses. It can help people through threatening situations, even when the hand-holding comes from a stranger. Be that as it may, relationship bonds can impact outcomes of hand-holding: Some research shows that stronger relationship bonds contribute to greater benefits for hand-holding.

Massage can provide several benefits for health and well-being throughout our life span. In particular, the benefits of moderate-pressure massage therapy have been shown across various groups, including cancer patients, those with chronic pain, and low-birth-weight infants. It is wise for those with health conditions to consult a physician before engaging in regular massage therapy and for everyone to work with trained therapists.

CHAPTER 4

Touch Hunger: What Happens When We Don't Receive Enough Touch?

Following the fall of communism in Romania in the late 1980s, news began to emerge of thousands of children who had been left to the state's care due to poverty and the strictly controlled fertility regulations of the regime. Former dictator Nicolae Ceaușescu had wanted to raise the country's industrial output by increasing the population. He introduced policies that restricted contraceptives, banned abortions for women who had not had at least four children, and placed an income tax on childless adults over 25. As a result of these policies, birth rates in Romania rose by 13 percent in a single year. This population explosion inadvertently and tragically initiated one of the saddest natural experiments on touch starvation.

By 1990, an estimated 100,000 infants and children were abandoned in orphanages across Romania. Many of these children were kept in understaffed institutions, segregated by age group, and received limited contact with others. They experienced extreme tactile and social deprivation in early childhood when they needed it most (as we saw in the previous chapters).

The consequences of this tactile deprivation were severe. They included unusual brain development, attachment disorders, and lower performance on cognitive assessments. One study even found that the cognitive deficits linked to the deprivation experienced by the orphans in early childhood persisted even after the children had spent over seven years in adopted homes.[1] Thankfully, situations like this are rare, but these troubling cases show us how important touch can be to our development. The consequences can be dramatic and long-lasting when we are deprived of it.

THE STORY OF HARLOW'S MONKEYS

By this point, we have seen several examples of how touch is an important sense for social connection. Still, we cannot attribute all the changes found in extreme situations of social deprivation—like that of the Romanian orphans—to a single source of social input. This is because the level of deprivation experienced in these extreme cases extends beyond tactile sensory deprivation. For instance, these children were known to also experience limited verbal interaction, limited opportunity for play, and impoverished living conditions.

There is, however, broader research showing the impact of touch deprivation on our development, health, and well-being. One example comes from the work of Harry Harlow, an American psychologist based at the University of Wisconsin who conducted controversial studies from the 1930s through the 1960s that involved separating young rhesus macaque monkeys from their mothers.[2]

Widely considered cruel and unethical in the modern day, Harlow's experiments make for uncomfortable reading. The research involved the infant monkeys being taken from their mothers just moments after birth and placed in cages with surrogate mothers.

One surrogate mother was made of nothing more than wire and wood. The other was made of the same materials but with foam and a soft cloth wrapped around it. The wire mother could provide food through a bottle connected to it. The soft-cloth mother could not. The baby monkeys were forced to choose between comforting touch and food.

Harlow wanted to see which mother the monkeys spent more time with: the mother who provided food or the mother who could give the faintest amount of tactile feedback.

The newborn monkeys chose the soft-cloth mother most often, showing a preference for tactile contact rather than simply food.

Harlow also wanted to see what would happen if the monkeys were exposed to something scary—a frightening object placed in the cage. Again, the monkeys went to the soft-cloth mother rather than the wire mother that gave them food. A mother that provided some form of comforting physical contact resulted in more security for the infant monkeys.

Sadly, the research did not stop there.

A new experiment was conducted. Once again, baby monkeys were taken from their mothers and put in cages. However, rather than having access to two mothers, they only had access to one: They were placed in a cage with either the wire mother or the soft-cloth mother.

Both groups received food, but how they digested it differed. The monkeys reared with a wire mother had difficulty digesting the milk. Harlow interpreted this as a physiological manifestation of stress due to a lack of tactile comfort. In his words, the wire mother was "biologically adequate but psychologically inept."

While Harlow's studies can be hard to reconcile ethically, I think we must learn from them. Those poor monkeys went through so much distress that we owe it to them at least to recognize that the results are a powerful demonstration of why touch is important. Depriving one of our closest relatives of touch during early life had severe and catastrophic consequences for social connection, behavior, and well-being.

Thankfully, studies like Harlow's are hard to find. Yet other sources of evidence indicate that lack of touch during childhood can impact behavior. Inspired by Harlow's work on macaques, the American developmental psychologist James W. Prescott was one of the first to draw links between a lack of tactile affection, such as holding and carrying, in early childhood and a propensity towards violence.

This work, reported by Prescott in 1975, investigated data from cultural anthropologists who had previously recorded observations from 400 different cultures worldwide. Some cultural groups studied included the Maasai people, the Māori people, and the Navajo people.[3]

Prescott examined the relationship between physical affection (such as caressing and playing with infants) and violence (such as frequency of theft) in 49 of the cultures. In most cultures, he found that high physical affection in young children was linked to low adult physical violence. Put simply, access to more touch was connected to less violence.

Prescott's findings are interesting but do contain methodological concerns. These include the fact that affection and violence were not measured in the same person, so no causal link can be made between the two factors.

Evidence hinting at a similar conclusion has, however, been provided. In 1999, Tiffany Field from the University of Miami examined interactions between adolescent friends at McDonald's restaurants in Paris and Miami. Field investigated how much touching took place within each group and how this related to verbally and physically aggressive behaviors. The results showed that American teenagers touched other people less and were more aggressive towards others than their French counterparts.[4] So again, aligning with Prescott, we can see a relationship between the amount of touch and aggressive behaviors.

We are still learning why this may be the case. As with Prescott's findings from 1975, we must be mindful that just because two things are related does not necessarily mean they are causally linked. In other words, we can't determine a causal link between less touch and more aggressive behavior from this work alone.

Field herself acknowledged that factors other than touch might contribute to why she found relationships between touch and aggression in American children and teens and their French counterparts. These could include differences in cultural norms. Or perhaps more aggressive behavior is a trigger for lack of touch, rather than reduced touch driving aggression. That is to say, aggressive situations may lead to less touch being observed.

Another possibility is that the groups may have differed in a variable that was not captured. As we'll learn later, people can vary in touch behaviors for many reasons, including their levels of attachment or personality traits. These were not measured in the studies, so we simply can't determine how any of these factors could be contributing to the pattern of results.

Whatever the precise mechanism of why, the combined results of Prescott and Field make a case for a relationship between the amount of touch in development and aggressive behavior.

FROM TOUCH DEPRIVATION TO LONGING FOR TOUCH

Studies on touch deprivation during early life show how wide-ranging and long-lasting the effects of being starved of touch can be. They also leave me wondering what happens if we are *not* touch-deprived in early life, but simply feel that we receive less touch than we want or need in later life.

Much of the work discussed so far in this chapter focuses on situations where there is an extreme and, thankfully, quite rare lack of touch. But there are situations where, as adults, we may receive some touch but still feel that we do not get enough quality touch to fulfill our needs. Can we be touch hungry in adulthood? What impact does this have on our health and well-being?

It turns out I was not alone in wondering about the impact of touch hunger in adulthood. In 2014, Kory Floyd from the University of Arizona published an article that directly addressed this question. Floyd surveyed 509 adults from 50 US states, the District of Columbia, Puerto Rico, and 16 other countries about their feelings of wanting more affectionate touch in their life. He also sought to see how these feelings of not receiving sufficient affectionate touch were related to social and health outcomes—including well-being, loneliness, stress, depression, anxiety, and attachment style.

Floyd found that affective touch hunger in adulthood was linked to various physical and mental health outcomes. People who reported wanting more touch than they received showed higher levels of loneliness, depression, stress, and mood and anxiety disorders. The same individuals who reported a stronger desire for more touch also had more difficulty expressing and interpreting their emotions. They showed lower general health, happiness, social support, and relationship satisfaction.[5]

These findings contributed to Floyd later explaining in his book *The Loneliness Cure* that meeting our desire for affectionate behaviors such as touch is linked to something we require for survival. According to Floyd, affection deprivation hurts when we experience it, and ignoring it can lead to health problems.[6]

To me, Floyd's work was a wake-up call. If we think of touch hunger as the difference between the touch we get and the touch we need, then any one of us could face touch hunger at some point in our life. And this touch hunger could impact both our physical and mental health. We don't need to be extremely deprived of touch, like Harlow's monkeys, to feel the impact of a lack of touch on our well-being. It led me to wonder how many people were feeling touch hungry right now.

One part of the Touch Test survey we conducted at the start of 2020 asked people whether they received enough touch in their lives. We found that 54 percent of surveyed people reported getting too little touch. In contrast, only 4 percent of people reported having too much.*

Some world regions reported feelings of touch hunger more than others, with almost 72 percent of survey respondents from North America and northern Europe reporting that they got too little touch—these were the most touch-hungry regions in our study.†

These findings reminded me of another piece of research reported in 2015 by Kory Floyd on "affection hunger"—the hunger for more intimacy and affection of any kind in our lives, rather than touch alone. In a survey of 1,500 Americans, Floyd found that 75 percent of adults agreed that "Americans are in a state of affection hunger."[7]

Let's pause for a second and let those numbers sink in. Over half of a world sample felt they had too little touch in their lives in 2020. Three-quarters of Americans surveyed felt that their country was in a state of affection hunger when asked in 2015. Findings like these have left some to question if we are living in a crisis of touch.[8]

THE GREAT TOUCH EXPERIMENT

I had always thought of myself as quite reserved when it came to intentionally touching other people, keeping my tactile behaviors for those close to me. I felt I engaged in touch a reasonably

---

* The remainder of the sample reported that they received just the right amount of touch.
† In contrast, the least amount of touch hunger was reported by respondents from Southeast Asia, where 45 percent of people reported that they didn't get enough touch in their life.

average amount. I would touch in settings that you might expect, like hugging my best friend as a greeting or holding their newborn child. I rarely touched strangers (the Free Hug in Sydney being an exception) and occasionally would touch colleagues.

Even so, living during the first COVID-19 national lockdown in the UK in March 2020, I became acutely aware of just how much I was missing touching other people. I was restricted to leaving my home for essential purposes only and had to follow strict social-distancing guidelines, including no touching. It wasn't necessarily the frequency of touch I was looking for, but a sense of quality human tactile interaction from friends and family.

I certainly wasn't alone. In a way, the national lockdowns of the COVID-19 pandemic made us all live through a big touch experiment. An experiment that was not just limited to the people who might participate in research in a lab on a university campus, but captured most of the world.

From research findings to a wealth of social media posts, both during and coming out of lockdown restrictions, people reported noticing a lack of touch.

In one study, Tiffany Field and her colleagues explored feelings of touch deprivation in American adults during the lockdown in April 2020. They found that 60 percent of the people surveyed reported experiencing low to high levels of touch deprivation. The degree of deprivation experienced was also associated with more negative mood and sleep disturbances.[9]

Separate research conducted between April 5 and October 8, 2020, found that almost 83 percent of adults reported wishing for more touch than they had in their lives at that time.[10] This research, led by Larissa Meijer and colleagues from Utrecht University, involved a sample of just under 2,000 adults mainly located in Europe. On average, the participants indicated that COVID-19 restrictions had been in place in their country for close to 43 days at the time of testing. Feelings of longing for touch increased with the number of days those regulations were in place, a result mirrored in other research showing that feelings of intimate touch deprivation during COVID-19 pandemic restrictions were connected to greater loneliness and anxiety in some people.[11]

So, like Kory Floyd's work before the pandemic, it seems that the need for more affectionate touch in our lives is linked with adverse outcomes for our mental health. When coupled with reports of high levels of touch deprivation during the social isolation period of the pandemic, these conclusions were sobering to me.

It's easy to imagine that some people may have been affected by touch hunger more than others during the COVID-19 lockdowns. This was borne out in the data. Those with a more anxious attachment style were more likely to report wanting touch during lockdown, particularly in the form of hugs and high-fives from friends. This makes sense, because anxious attachment styles are often associated with craving closeness and reassurance from others.

To help explain this further, it is helpful for me to introduce you to what we mean by attachment styles. Attachment styles are most straightforwardly explained as differences in how we interact and behave in our relationships—not just romantic relationships but also those with friends, family members, etc. In short, attachment can influence how we engage in behaviors related to proximity and support-seeking from various people throughout our lives.

It's been found that each of us has individual differences in how we think and feel about attachment—our attachment style.[12] For instance, some people may seek out intimacy while others fear it. Some people may prefer to be more independent from their partners than others. A balance of these behaviors can support a secure attachment style—where people feel confident and comfortable in their relationships.

Yet sometimes people can show substantially more of one attachment style than others. One example is characterized by showing more attachment avoidance: the tendency to be more likely to fear intimacy and show greater independence from a partner. To help picture this, fans of the TV show *Gilmore Girls* may think of the character Lorelai, who literally runs away from her wedding and repeatedly avoids intimate emotional relationships with her partners.

In contrast, some people have a more anxious attachment style: To use *Gilmore Girls* to help clarify again, we might consider the character Dean, who dates Lorelai's daughter Rory. In Season 2, this relationship starts to go south. The arrival of a new man in Rory's life leads to jealousy and insecure feelings in Dean. He becomes clingy and untrusting.

During the COVID-19 pandemic, people with a more anxious attachment style were more sensitive to reductions of touch in their life. In contrast, people with higher attachment avoidance were less likely to report wanting to experience touch, particularly intimate touch. This is understandable, as those with higher attachment avoidance prefer to maintain independence and emotional distance.

These results show that in adults, as in infants, touch hunger can lead to negative outcomes for well-being. But we must be wary of assuming a simple relationship of more touch equaling a better outcome for everyone.

Humans are complex individuals and not everybody wants more touch. Some people may be more comfortable with reduced touch. Our needs will differ, as will our outcomes. When it comes to touch, we should always take other people's personal preferences into account. A one-size-fits-all approach just won't do. It is not a simple case of more is better or less is worse. Rather, it is about matching the quality of touch each person receives to their own desires.

TOUCH HUNGER IN VULNERABLE ADULT AGE GROUPS

The fact that experiences of touch hunger during lockdown varied according to attachment style left me wondering what other factors might influence how people respond to reduced touch. While I missed touch during the national lockdowns, I was lucky to be living with my partner and could still experience affective touch. Yes, I missed supportive touch from friends and family, but I still had touch in my life. A daily hug. A reassuring arm on my shoulder. A partner to share affection with day to day.

What about someone who lived alone? Or young adults separated from family and friends? What about people in nursing homes who were not allowed to see, let alone physically hug, a loved one? Thinking of the many ways the pandemic affected individuals, I wondered who might be especially vulnerable to touch hunger in our society.

Some data spoke to these questions. In Tiffany Field's survey of American adults conducted in April 2020, touch hunger was more common in people living alone. In the work by Larissa Meijer with predominantly European adults, people who lived alone were more touch-deprived than people who lived with housemates, while women and nonbinary adults reported longing for touch more than men.

In my journey to answer these questions further, I spoke with a friend who had been unable to see her father throughout the lockdowns because of restrictions limiting visits to nursing homes. "It's been one of the most distressing times," Sarah said to me as we spoke over Zoom. "Dad has dementia. He has problems with his vision and his hearing. Touch is our main source of connection," she continued. "We've been forced to watch from afar as he declines further. Not being able to touch him or to help him understand why. I last held his hand months ago. The closest I've been able to get is meters away. It must be so frightening for him."

Sarah's description of her father's isolation and its consequences on his health was harrowing to hear. The decline that she described is consistent with evidence showing how a lack of social contact can relate to cognitive decline. To give you an idea, one study from the English Longitudinal Study of Ageing investigated how social isolation impacted memory function in more than 11,000 people.[13] It found that people who reported higher-than-average social isolation also experienced above-average declines in memory function over a two-year period.*

"I really hope that when we can hold hands again; it will help," Sarah told me. "I want to help provide him with some comfort, to be able to communicate with him in the way we used to. Touch was so important to that."

Sarah's words reminded me of a phrase that has been mentioned to me regularly: "Touch is the first sense developed, and perhaps the very last to go." As we now know, affective touch responses continue to increase in humans until they at least 90 years old, if not longer.

Touch is vital to communication throughout our life span, playing a role in communicating, nurturing, and helping each other. Yet even before the pandemic, there were some reports that half a million older people in the UK can go at least five days a week without touching anyone.[14]

People living alone reported wanting more touch during the pandemic than people who were cohabiting. Generally, living alone is also known to put people at greater risk of loneliness.

---

* It should be noted that we do not know the direction of the relationship between social connection and memory. It could be that more social connection protects against memory decline, but it could also be the case that people with better memory are more likely to maintain stronger social connections.

When you consider that older adults are also more likely to live alone—at least in countries in the Western world—touch, or the lack of it, is a vital consideration in the elderly.

Of course, older adults are not the only group at risk of experiencing a lack of touch in their lives. Going back to Floyd's 2014 study on affective touch deprivation, it turns out that there is no correlation between age and wanting more affectionate touch in life. In other words, touch hunger affects people of all ages. That is not to say that older people do not want more touch in their life—some do—but put simply, the data suggests that on average, older people do not feel they lack touch any more than younger people do.

The lack of effects of age on feelings of touch hunger is perhaps not as surprising as it may at first seem. Older people are often at an increased risk of loneliness, but so are younger adults. This is especially true for young adults aged 18 to 25 years, who are often identified as one of the loneliest groups in society.[15]

Some estimates suggest that one in three people aged 18 to 25 years report feeling lonely three or more times a week. We do not know all the reasons for this, but it has been suggested that psychosocial factors may play a role. This time of life is filled with experiences that could contribute to loneliness, such as feeling homesick after making the first move away from home, feeling disconnected from social groups or communities at a time when major life transitions are taking place (e.g., leaving school), and the stresses around making critical decisions linked to career and personal life.

During the pandemic, 63 percent of young adults reported symptoms of anxiety and depression.[16] This is quite an unsettling statistic when combined with their risk of loneliness. The sharp reduction in their ability to interact with friends and family during the pandemic hit them hard.

Many may have been single or living away from support networks—a challenging cocktail of risk for feelings of touch hunger and social disconnection. Like the elderly, young adults felt the effects of pandemic-related lockdowns on their mental health and touch hunger.

"My mental health had never been worse," said Jane, a university student who spent much of the pandemic-related lockdowns living away from home with two roommates under stay-at-home orders. "The lockdown felt like existing, not living," she continued. "It is one thing being away from family if you have friends you can meet and greet and interact with in a normal way. But when this stopped without notification, I missed the connection with other people. Don't get me wrong, we could still talk virtually, but I used to see my friends daily, and I missed being able to physically interact with them—to touch them. It may sound silly that not being able to simply hug a friend or laugh and put a hand on their arm mattered so much, but it really did."

We know that the amount of touch we desire is a very personal thing. But adapting from a norm where touch was available to a situation where it was withdrawn overnight was challenging for everyone. Jane's account resonated with what I heard from many university students living away from home during the lockdowns. Students everywhere were struggling.

Her comments also reminded me of some of the steps people took to overcome the lack of touch when they were home during lockdown. In one of the Touch Test broadcasts, we interviewed B and Z.[17] They were friends who lived with a group of six people in a rented house. When lockdown hit, they were the only people in the house, and were unable to see partners or friends. Their social connection was changed overnight. To help overcome feelings of anxiety and a lack of touch, they came up with the idea of a safe daily hug. Initially starting as a joke between them, it soon became routine and ultimately necessary, helping them through a time when many people were longing for more touch in their lives.

THE CUDDLE THERAPIST

The story about setting time aside for a hug with a housemate during lockdown reminded me of a growing industry—cuddle therapy. Cuddle therapy isn't quite the same as a situation where two friends have a daily hug, because it involves two strangers. But the premise is not too dissimilar. It involves people paying for a cuddle from a trained cuddle therapist.[18]

In a typical cuddle therapy session, there is some form of consultation establishing a prior relationship with touch between therapist and client. As we've noted, touch is highly personal, so this consultation might include discussions around where the client would like to be touched (or not), their feelings about touch in their life, and any stressors that are currently present. From there, a variety of options are offered, such as hand-holding, massage, caress, hair-playing, or spooning.

Consent is essential throughout cuddle therapy, every step being checked between therapist and client. Sessions are also always fully clothed, with no sexual touching allowed. Care is taken to ensure that the client is not booked with the same cuddle therapist too often, in order to avoid the risk of overreliance and emotional bonds forming.

Perhaps surprisingly, there is also no requirement to touch during cuddle therapy if the client does not wish to do so at first. This is important because of nuances in touch preferences that vary from person to person. Some people may feel comfortable touching a professional straight away. In contrast, others may need time to build more rapport before doing so. In a world where people are increasingly reporting that they don't get enough touch in their lives, cuddle therapy is one option to help to bring touch to those who crave it.

Like many of us, cuddle therapists had to adapt their services during the COVID-19 pandemic. Many companies moved from face-to-face cuddling to virtual cuddle sessions. My curiosity (and possibly my feelings of a lack of touch in my life at the time) led me to see what might be involved in a virtual cuddle therapy session.

I was aware that, given how new the approach is, research findings on the effectiveness of hugging therapy are hard to find. Yet I tried to stay open-minded, while admittedly apprehensive about what might happen. Would the session draw on skills that have already been linked to well-being? How would cuddling actually happen without someone physically there? I was soon put at ease.

Freya, the cuddle therapist leading the call, started with some ice-breaking questions. A discussion about stresses in my life followed, and an open conversation about how touch might be missing. Freya emphasized the naturalness of feeling a lack of touch—even though some touch may be present, it is still possible for people to feel touch-deprived.

We went on to discuss memories of a cuddle that felt good in the past. I was asked to imagine how it felt both physically and emotionally, to focus on how my body felt then and how it was feeling right now.

The tactile imagery that Freya asked me to use reminded me of results from the Touch Test that suggest people with better tactile imagery show higher levels of well-being. The focus on how my body felt in the moment was almost meditative. Being mindful of specific body parts helped me stay present with Freya's guidance. This made me think of the evidence that mindfulness meditation can help self-awareness, well-being, and even physical health.

Although I am not aware of data directly investigating factors that may or may not contribute to the effectiveness of cuddle therapy, I could see how the use of positive tactile imagery and body-focused awareness could potentially be beneficial. And perhaps most importantly, the session made me feel more relaxed and positive.

Cuddle therapy is one route that people might take to tackle feelings of a lack of touch in their life, but are there others? Another closely related trend is cuddle parties, a mixture of workshop and social gathering where people are encouraged to talk openly about the desire for touch and to engage in nonsexual touching like cuddling.

There have even been cuddle cafés. In 2015, the biscuit brand McVitie's launched a pop-up cuddle café in London to help people make up for lack of touch in their lives. It allowed visitors to enjoy tea, cake, and biscuits for the price of one cuddle.[19]

Of course, cuddling people you do not know may not be everyone's cup of tea. Are there other touch substitutes that people can use? It turns out that quite a few different approaches are being developed to try to help those experiencing touch hunger.

On a visit to Japan, I came across a product called the tranquility chair,[20] built in the shape of a soft doll. It had long arms that could wrap around the person sitting in it, engulfing them almost like a bear hug. I have to admit that I didn't try it, but clearly, cuddle therapy is not the only approach seeking to address increasing reports of touch hunger. In fact, the makers of the tranquility chair indicate that one reason for building the product was to support attempts to reduce loneliness in the elderly and those who live alone by introducing comforting touch.

Many other technological devices are being developed and marketed to reduce touch hunger and loneliness. This has particularly been the case in the emerging field of social robotics, which is bringing science fiction closer to reality by building sophisticated artificial robots that can interact with people.

Take Qoobo,[21] a "tailed cushion that heals your heart." Qoobo is an example of therapeutic robotics. When it is caressed, its tail waves gently. When rubbed, the tail swings playfully. When dormant, the tail occasionally wags to say hello. By offering people touch and interaction via a social robot, the developers aim to provide a product to bring comfort, communication, and touch into the daily lives of those who suffer from touch hunger.

If you like the idea of a tactile cushion but are put off by the wagging tail, you might be interested to hear about the Calming Cushion. In 2022, researchers from the University of Bristol developed a huggable textured cushion that simulates slow breathing. It achieves this by using a controllable pneumatic chamber so that the cushion gradually moves when held. These movements mimic the type of slow breathing rate that can be helpful for soothing anxiety.

The reason for using a tactile cushion to mimic slow breathing was intentional. The researchers wanted to establish whether holding a cushion that breathed slowly might help people's anxiety levels. They found this to be the case: Using the cushion was effective for reducing anxiety in students who were going to undertake a math test compared to a no-intervention condition. The Calming Cushion was also just as good as guided meditation in helping to reduce anxiety levels in the students.[22] It's a wonderful development for those who crave some tactile comfort in their life, with the potential to help anxiety as well.

Aside from robots and breathing cushions, a more common approach that people often report as a touch substitute is stroking pets. Research has demonstrated that people show reduced blood pressure and decreased stress while petting their animals. In one study, scientists from Washington State University found that just 10 minutes of petting a cat or dog led to reductions in the body's primary stress hormone, cortisol.[23] This explains why stroking a dog or a cat can often make us feel more relaxed. It hints at why people turn to pets as a substitute for lack of touch in their lives.*

Even viewing other people stroking pets can substitute for a lack of touch in our lives. This experience is called vicarious touch; in other words, sharing the tactile experiences of other people by watching them.

When we watch other people experiencing touch, we tend to recruit similar parts of our brain as when we experience touch ourselves.[24] Let's say that I saw you being touched on the neck; the parts of my brain that become active when I feel touch to my neck would become active just by watching you being touched. This vicarious tactile response doesn't lead to me feeling what you are feeling, but it mirrors your experience.†

The fact that we can share the tactile experiences of others in this way might explain why many people are drawn to using observed touch as a touch substitute. To illustrate, are there any videos of people touching on YouTube or TikTok that you happen to find enjoyable or relaxing? Perhaps you might think of watching videos of people stroking their pets or being stroked themselves. Anecdotally, people frequently report using these as a touch substitute. Some research has also shown that the more time people spend longing for touch, the more pleasant they find watching videos of other people's arms being stroked at CT-optimal speeds.

Throughout the COVID-19 lockdowns, there were also online trends connected to videos of people cuddling. These videos were not always highlighted as a touch substitute, but vicarious touch helps to explain how we can share in heartwarming trends like the #CuddleChallenge. This trend involved children watching

---

* This can also include engaging with pets that people do not own. For instance, some organizations regularly put calls out for cat cuddlers.
† I should note here that there are some people who do feel what they see, but we'll come back to them in Chapter 7.

their favorite movie or TV show. Partway through, the parent lies next to them and puts their head on the child's lap. The outcome is often a lovely cuddle, but sometimes it's a hilarious rejection. I recommend a look if you want a brief bit of entertainment—laughter can also be good for your wellness.

Another touch substitute that we've already encountered is touch in treatment settings. For instance, a massage, a haircut, or a manicure. You may not immediately think of these as behaviors that people engage in to simply experience touch. Still, for some people these situations offer the chance to experience touch in a safe way.

Touch substitutes also don't necessarily have to be socially focused. It has been reported that activities like taking a hot bath or swimming are ways to introduce touch that is missing from life.

"I like warmth, and I will spend a fair amount of time, when I don't feel right, in my bathroom. I think of the bathroom as a place where we get in touch with our skin, which is our largest sensory organ; it's a great place to figure out different ways of being with your sensory system," says Vera, a dancer, choreographer, and researcher who I met during a podcast that we recently worked on together.[25] We discussed touch hunger and Vera's own work to reduce it. Her comment about baths aligned with broader data showing that people reported taking more baths as a touch substitute during the coronavirus-related lockdowns. I wondered if there were other types of touch substitutes that Vera used.

"Was there anything you did during lockdown to reduce touch hunger?" I asked.

"I basically started to lay out different fabrics or carpets or pieces of fake grass that I have at home in different locations around the apartment. It's just something I did, not thinking much around it, but in hindsight, it was definitely a way of keeping connected to this sense that I've explored in so many ways over the years," she told me.

There are other exercises that Vera described to deal with her need for sensory stimulation. "Okay, for this exercise, you need an object, ideally a stone, like, as big as your hand. Put that stone in front of you and take a seat. Maybe you can shift a little bit from one side to the other. You can also tilt a bit back and front and arrive with your spine in a sort of central position. Feel free to close your eyes. Have your hands resting on your knees. Now reach for your object. Take it in both hands and start exploring it. You can slide your fingers over it. What are you feeling? Is it maybe uneven in parts? Porous? Bumpy? Does it have any edges? Are the edges round? Are there different textures on different parts? What is there to say about the quality of this object? Can you press into it? Is it maybe hard? Solid? Is there maybe a part that is very different to the rest? Is it the quality? The texture? The surface? And now, since you've been exploring for a while, maybe observe your own hands a little bit. Is there a bias? Does one hand explore more than the other? Are you exploring with both at the same time? What about the temperature of the object? Is it different to your own body temperature? Or maybe in time with the exploration, the object is warming up."

I really connected with Vera's idea: combining touch with nature to help people connect with their bodies and the world. It also reminded me how many people who work with natural materials such as clay report tactile gratification and restorative benefits from engaging with ceramic materials. While often a solitary activity, creating with touch through clay has a long history of self-reported benefits to health and subjective well-being.

These simple and low-cost approaches do not require technology or quirks. They simply connect people to natural materials through touch. Clay, a leaf, a stone, or a twig provides powerful ways to experience tactile gratification from materials that many can access with ease.

Ten accessible ways to tackle touch hunger

It's clear that even if we live with others, everyone can struggle at times with the experience of touch hunger. Many people report not getting enough quality touch in their lives. We know that the amount of touch we desire is a very personal thing, but there is no doubt that it is an integral part of how connected we feel with other people. There is widespread consensus that social connection protects and promotes mental health, while people with stronger social relationships have also been shown to live longer. Knowing the importance of touch for health and loneliness, we must consider how we can support those who crave more touch in their lives—perhaps you are one of these people. Of course, touch substitutes can't solve all the complexities of being lonely, but they may take the edge off.

Here are ten accessible ways you can bring the benefits of touch into your everyday life.

1. Take a warm bath.* This has the effect of improving blood flow, cleansing the skin, and regulating how your body feels. Bathing has also been linked to lower stress.
2. Learn to dance. This can provide a source of exercise and social connection, and a chance to experience shared touch, since many dances are built around human contact.
3. Engage in tactile self-care treatments like a massage or getting your nails done. This provides a source of touch from other people and can help with relaxation. Massage in particular has been shown to be effective in supporting

---

* Be mindful that high-temperature baths can put unnecessary strain on your heart, especially if you have a preexisting heart condition. In general, we should look to use warm rather than hot water.

sleep, reducing pain, and improving aspects of physical and mental health.
4. Stroke a pet or an animal at a petting zoo or pet café, sign up at a dog-walking site, or volunteer as a cat cuddler. This can help reduce stress levels, build social connections, and offer a route to physical exercise, all of which carry benefits for physical and mental health.
5. Explore the feel of different objects—an item of clothing, perhaps, or anything you like to touch. This may help develop our sensory skills. By focusing on how different textures feel and how we respond, we can also become more mindful of our bodies. Different textures can trigger memories of positive touch that can help us imagine situations where touch felt good.
6. Cuddle an inanimate object like a cushion or a pillow. This provides a source of sensory input. It may also help provide an environmental cue to trigger previous states, such as feeling secure following a hug or cuddling to relax before bed.
7. Talk openly about touch. This offers the chance to help other people understand your touch needs. Touch hunger can happen when the touch we receive doesn't match what we need, but people's preferences vary a lot, and it can be challenging for us to understand what others need without clear guidance. By communicating, we can help people understand what works for each of us.
8. Provide people with opportunities to touch you. This can give the time and space that may be needed for touch in your life. Giving positive touch to someone else can help them and offer the opportunity for you to receive touch in return. Even something like greeting people with a hug may work if this type of greeting is appropriate for the person you meet.

9. Consider cuddle therapists or cuddle cafés in your area or online. This can provide a safe and structured space to talk openly about touch, recall positive touch experiences, and/or experience touch.
10. Watch other people or pets being touched via online videos. This has been reported to reduce anxiety and act as a touch substitute. It may be a good alternative for people who have few opportunities for touch with other people around them, or even a good addition for those who do.

CHAPTER 5

Tactile Intimacy: From Sex to Spooning, the Role of Intimate Touch in Relationships

To get us started in the final chapter of this section, let me ask you a question. Have you ever seen the movie *Never Been Kissed*?

I have to admit that I have a bit of a guilty pleasure for romcoms. Whether it is *500 Days of Summer*, *10 Things I Hate About You*, *Legally Blonde*, *Legally Blonde the Musical*—you name it, I've likely seen it. One of my favorites is *Never Been Kissed*, in which the heroine, Josie Geller, describes a kiss between two people as "that thing, that moment . . . this amazing gift."

I'm sharing this not just to demonstrate my encyclopedic knowledge of late-nineties romantic comedies, but to get us to pause and think about memorable touch experiences, particularly with romantic partners. As we've seen already, we share many types of touch throughout our lifetime. Sometimes with caregivers. Sometimes with friends. Often with romantic partners.

The forms of touch we engage in with these people can take very different meanings and routes depending on the nature of our relationship. Let's take kissing as an example. Kissing is an act that is reported to happen in up to 90 percent of cultures. It can be a greeting between friends or family members—a peck on the cheek or a kiss goodnight. It can also have a more romantic and sexual meaning when dating someone or throughout different stages of a long-term romantic relationship. The volume of different types of kisses people engage in has led many to report on the possibly ubiquitous nature across human cultures.

Yet when we look more closely, some types of kissing are less common than we might think. One study, led by researchers from the University of Nevada, Las Vegas, examined the presence of romantic–sexual kissing in 168 cultures. They defined a romantic–sexual kiss as lip-to-lip contact between lovers that may or may not be prolonged.

The researchers searched databases and historical sources for references to different types of kisses across the regions studied: Africa, Asia, Europe, Middle America, the Caribbean, the Middle East, North America, Oceania, and South America. They also asked ethnographers who had visited regions within the study: "Did you observe or hear of people kissing on the mouth in a sexual, intimate setting?"

By using these different sources of knowledge and carefully checking for comments about the presence of other types of kissing (e.g., as a greeting between friends or family members), the research team recorded whether romantic–sexual kissing was present or absent across the 168 cultures under investigation.

What they found surprised me. Only 46 percent of the cultural areas studied engaged in lip-to-lip kissing in a romantic–sexual sense.[1]

There were some subtleties to this around the world: In North America, romantic–sexual kissing was present in 55 percent of the cultural regions, while in Europe, that was true for 70 percent of cultural areas. Still, having grown up surrounded by a popular culture where romantic kissing is so commonplace, I was surprised that fewer than half of the cultural areas studied appeared to engage in this behavior.

Why might this be? There are natural limitations to the study—just because romantic–sexual kissing was not observed or heard of doesn't mean that it doesn't take place. But even so, the low level of engagement in romantic–sexual kissing across all cultural areas investigated seems a far cry from those Hollywood love stories in which the anticipated and iconic kiss between romantic partners is the moment you are waiting for.

My surprise may be because I live in a cultural area where romantic–sexual kissing is quite prevalent. In the same research, a strong connection was found between the frequency of romantic–sexual kissing and the industrialized nature of a culture. The more industrialized a culture was, the more likely it was that romantic–sexual kissing would occur. That is not to say that more skyscrapers and heavy industry cause people to kiss more. Merely there is an association between the level of industrialization and the amount of romantic–sexual kissing: a correlation rather than causation. Romantic kissing may well be so prominent to me—or you—because we live in an environment that promotes it.

When you stop and think about it, romantic kissing is a strange concept. On the one hand, there are kisses like those described in movies, an incredible moment that connects you with a partner. Kissing can be exciting. It can be passionate. It can be amazing. On the other hand, kissing can also be nerve-racking—not necessarily in a bad way, but anxiety-inducing nonetheless. There are also, sometimes, those unfortunate situations where kissing can be a little bit . . . how can I put it? Awkward.

Do you remember your first romantic kiss? I wonder which of the words above you might use to describe it—amazing, exciting, passionate, nerve-racking, anxiety-inducing, awkward.

There's one description of romantic kissing that you rarely see people use, although it often occurs. Two people put their lips together and exchange saliva. While this rather takes the romance out of it, it is accurate. It's also quite a different way to think about kissing than the "amazing gift" description that we sometimes get from popular culture.

Lip-to-lip kissing involves the passing of bacteria and saliva between mouths. This can facilitate all sorts of exchanges we might not intend, such as passing viruses to one another. One study found that in 15- to 19-year-olds, romantic kissing quadrupled the risk of a form of bacterial meningitis called meningococcal meningitis.[2]

Not only that, but exposure to allergens can occur through exchanging saliva during romantic–sexual kissing. Take peanuts. One study investigated how long peanut allergens stay in the saliva following ingesting two tablespoons of peanut butter.[3] For some people, the allergen stops being present after five minutes, but it took hours for others. Given that peanut allergies can be passed on through sharing saliva, this means there are sometimes hours-long risk periods. In a world full of peanut butter fanatics—supposedly a third of us report peanut butter as the one food we couldn't live without—that's an unknown risk to kissing that we might not immediately think of.

ONE KISS IS ALL IT TAKES

Given the various health risks connected to kissing, you might be wondering why we kiss at all. References to kissing as a sign of affection can be traced back to Vedic Sanskrit texts, composed around 3,500 years ago, although many assume we've been kissing much longer. Part of the reason for this belief is that it is not just humans that kiss: Did you know that even bonobos and common chimpanzees kiss and embrace after fights?[4]

Discussions about the origin of kissing in humans often draw attention to the fact that, as babies, we appear to have a particular liking for lip touching. Babies are constantly touching and putting things near their lips. Case in point, I recall my friend's baby becoming a fan of self-foot-sucking once they reached six months.

There are several reasons for this. One reason suggested by some scientists is that babies may associate touch on their lips with the experience of receiving food. With this positive reinforcement, lip touching is linked to something good.

It has been suggested that the evolution of kissing in humans may have been built from this association. More specifically, evolutionary anthropologists have drawn attention to something called premasticated food transfer.[5] This refers to the notion that our ancient ancestors may have chewed food and transferred it directly into their babies' mouths, like birds feeding bugs to their chicks.

Arguments connecting kissing to premasticated food transfer go something like this: Mouth-to-mouth food transfer from our ancestors enabled the development of associations between pressing our lips together, caregiving, and a feeling of reward—originally food in this case. Over time, this association grew stronger and stronger.

Eventually, the strength of association between pressing our lips together and rewarding feeding experiences reached a point where mouth-to-mouth contact moved away from the act of feeding alone. It devolved into a more general expression of affection and care that started to be applied to other relationships. In other words, by repeatedly pairing our mouths together to share food, we began associating lip-to-lip contact with the naturally rewarding and caregiving eating experience. This association meant that we wanted to bring our lips together more.

Irrespective of the precise evolutionary origins of kissing, what is more apparent is that our lips are a very sensitive part of our body. There are a vast number of receptors in the lips. The tactile body maps in our brain also have a large representation of the lips compared to other body parts—you'll know this all too well if you looked up our somatosensory homunculus friend back on page 26.

All of this contributes to powerful sensations when our lips are touched, whether by someone tracing their finger around them or by our lips locking with someone else's. Their sensitive nature may contribute to why kissing feels good.

When considering why we engage in romantic kissing, it is also important to remember that kissing rarely occurs in isolation. When we romantically kiss, we don't just touch lips; we can often touch in other ways, such as hugging, hand-holding, or sex. This adds to the tactile experience that can be associated with romantic kissing. It also links kissing with other forms of tactile intimacy that can be important for our personal state and emotional bonds with other people.

We've already seen how forms of tactile intimacy like hugging and hand-holding can positively affect our health and well-being. Kissing is similar. One study of healthy adults in marital or cohabiting romantic relationships found that increasing the frequency of romantic kissing over six weeks improved relationship satisfaction and perceived stress. It even reduced serum cholesterol (which can be elevated by stress) in the couples involved in the research.[6]

Given this, it is not surprising to hear that both men and women rate kissing as one of the most highly romantic acts a couple can engage in. A study conducted by researchers at Brigham Young University found that kissing on the lips is rated by both male and female university students as more intimate than cuddling, holding hands, hugging, caressing, stroking, massages, back rubs, and kissing on the face. What's more, the amount of reported kissing between partners was connected to relationship satisfaction.[7] Far from being a simple exchange of saliva, romantic kissing is an important contributor to our relationships and, likely, in turn, to our well-being.

*

Kissing isn't just important to how we maintain our romantic relationships. It is important in how we establish them as well. In

2014, researchers from the University of Oxford published a study called "What's in a kiss? The effect of romantic kissing on mate desirability."[8] Across two experiments involving over 900 people, the researchers examined how our ability to kiss is used to judge potential partners. In other words, does it matter if you are a good kisser or not?

In one experiment, the research participants were given written dating profiles of potential partners. The profiles contained information about personality traits and the person's relationship history. Each profile had a similar number of positive and negative descriptors. Some hobbies, like cooking or running, were also mentioned.

The profiles also contained information about the potential partner's relationships, sexual abilities, and experiences. These included descriptions of how good previous partners thought they were in bed and how intimate they felt they were.

The participants in the study saw the same profile for each potential partner. There was, however, an important difference. Half of the participants were told that the person was a "good kisser," while the others were told that the person was a "bad kisser."

To make that a little clearer, imagine you and I were given two profiles, one for someone called River and another for someone called Jordan. Although we'd been given the same information, you'd be told that "River is a good kisser" while "Jordan is a bad kisser." In contrast, I'd be told "River is a bad kisser" while "Jordan is a good kisser." Same profiles, same people, but the experimenters had introduced a simple difference: whether the person was described as a good kisser or not.

After viewing the profiles, the research participants were asked how attractive they found the person described. They were asked to indicate their interest in going on a date or having casual sex with them. They were also asked about their interest in pursuing a long-term relationship with them.

The results showed that people described as good kissers were rated more attractive than bad kissers. The good kissers were more likely to get a date.

Participants were also more interested in having casual sex with them. The good kissers were more likely to get lucky.

Good kissers also received higher ratings for interest in pursuing a committed relationship. A hat trick of gains simply by being labeled as a good kisser.

Before you start practicing kissing your hand, fruit, or whatever else Google suggests as kissing training, the eagle-eyed among you may raise a reasonable objection to the results from the University of Oxford study. Sure, being described as a good kisser can impact how people interpret written dating profiles, but in typical dating settings, we usually see what the person looks like too.

To address this, the Oxford researchers conducted a second experiment. Here, they tested a new group of research participants. This group were given the same written profiles, but this time they were accompanied by pictures of the potential partner.

So, going back to our previous example, once again, you and I would be given the same information for Jordan and River, but this time including a picture for each of them. You'd be told, "River is a good kisser" while "Jordan is a bad kisser." I'd be told the opposite. Would being able to see the person substantially change the pattern of results?

No. A similar pattern emerged as in the original study. People were still more likely to indicate an interest in casual sexual encounters or long-term relationships if they believed the person was a good kisser. Although in this study, the ratings of attractiveness and likelihood of going on a date were not influenced by perceived kissing ability.

Merged, these findings show us that our perceptions about how good someone is at kissing can be a salient factor in partner assessment for both casual and long-term relationships.

IT'S THE WAY YOU LOVE ME

The evidence we've seen shows us how powerful romantic kissing can be, impacting partner selection, relationship satisfaction, and even stress. Of course, we must be mindful that how much significance we place on romantic kissing in relationship formation and maintenance will likely vary from person to person. People engage in kissing for various reasons. How we respond to a kiss depends on a host of individual differences.

One key area where individual differences in kissing have been studied is in the context of gender differences between men and women. I want to note here that our following paragraphs focus on men and women because, sadly, there is limited research on people reporting more diverse gender identities at present—another reminder of the need for more diverse representation in research studies.

We've already mentioned that men and women share the view that kissing is a romantic act. Yet this is only part of the story. Men and women choose to engage in kissing for different reasons.

One example is that men are more likely than women to report using kissing to end an argument. Both males and females say they do this, but males are more likely to turn to a kiss as a form of conflict resolution with a romantic partner. In this regard, we can think of kissing as a reconciliation gesture. It is a way to help maintain our bonds in relationships by reestablishing them after an argument. Men seem particularly sensitive to this.

Along with helping us make up after a fight, other gender-related differences have been reported. In one study, over 1,000 university students in America were asked about kissing preferences, attitudes, styles, and behaviors. The research, led by a team from Albright College in Pennsylvania, found that female university students reported kissing to be more important before, during, and after sexual intercourse with both long-term and casual partners than males.[9]

Male and female students also differed in how important they thought kissing was for initiating sex, with 53 percent of males indicating they would have sex without kissing. In comparison, only 15 percent of female students indicated they would consider having sex with someone without kissing them first.

Male students in the study also felt that kissing should lead to sex to the same degree with both a long-term and a short-term partner. In contrast, female students thought kissing should lead to sex more with a long-term partner than with a short-term partner.

Focusing more closely on long-term relationships, the researchers found differences in how male and female students think about the importance of kissing at the beginning and later in a relationship. Female students placed relatively constant importance on kissing throughout the relationship—from the very beginning to their current relationship status. Male students differed, reporting that the significance of kissing decreased as the relationship progressed over time.

As an aside, I have to admit that these findings don't align with my own views about the importance of forms of tactile affection throughout relationships. There may be some readers who agree or disagree with the thoughts expressed by the students in this study. This is a gentle reminder that just because one group of male and female university students felt one way doesn't mean that every male or female will feel the same way.

Notwithstanding my interjection in the last paragraph, we might ask why we see such pronounced differences between men and women in these research studies. Some researchers have suggested that kissing shares information regarding the suitability of a partner. To help explain this, I need to introduce you to the major histocompatibility complex (MHC).

The MHC contains a set of closely linked genes that form part of, and contribute to, an adaptive immune system.[10] Different people can have a similar or dissimilar set of MHC genes. For our purpose, what is important to know is that humans have been shown to prefer the natural scent—body odor—of potential partners with different MHC genes.

For instance, one study found that women like the smell of T-shirts worn by men with dissimilar MHC genes to them. An argument put forward to explain why this might happen is that when two people with different MHC genes mate, the baby they produce could have a more diverse immune system and, thus, a greater ability to fight disease.

Although a potentially helpful signal, there is a challenge in using natural scent in mating selection—it is not straightforward to detect. Some scientists suggest that romantic kissing can help bring us closer to one another to perceive cues that may be important to making judgments about mating. This includes smells and tastes, which can indicate another person's health.

Kissing also involves exchanging an oily substance from the surface around and inside the mouth, known as sebum, which some have suggested may carry pheromonal and hormonal information. In other words, we might unknowingly use smell and taste during kissing to judge mate quality.

So what does this have to do with gender differences in kissing?

Interestingly, gender difference exists in the perception of smell and taste, with women having a heightened acuity compared to men.[11] This acuity difference shifts over the course of the menstrual cycle, becoming more pronounced during ovulation. It has been argued that these differences in sensitivity may explain why women place more importance on kissing than males when assessing partners. The argument is based on the assumption that women can use kissing as a more powerful way to detect differences in smell and taste than men.*

Whatever the reason for gender differences in the thoughts about the role of romantic kissing in relationships, there is little question that romantic kissing is an important nonverbal behavior. It is a rich and complex tactile embrace that people engage in for many reasons. To express affection, to show attraction, to support bonds. And, of course, to increase sexual arousal.

THE NEUROSCIENTIST'S GUIDE TO SEXUAL TOUCH

Before we get going in this section, it is worth mentioning that despite the provocative subheading, I am no Carrie Bradshaw. I could only wish to have the writing skills of a sex columnist like her—credit, of course, to Candace Bushnell, who published the book *Sex and the City* based on her own columns.

In a chapter discussing consensual intimate touch between romantic partners, I would be remiss, however, not to discuss sex.

So far in this book, I've explained that we have a high number of touch receptors in sensitive areas of our body, like our hands and lips. You'd be right to guess that we have many nerve endings in more intimate regions of the body, like the genitals, too.

---

* It should be noted that this is not to say that men do not use chemical cues during kissing. Several changes occur throughout the menstrual cycle that might be detected during kissing, including factors contributing to differences in breath and odorless molecules in saliva while ovulating.

In his 2005 book, *Touch*, author David Linden provides a detailed description of the biology of sexual touch.[12] He rightly notes that, like the perception of all sensory sensations, how we perceive touch in the buildup to or during sex is connected to our expectations and context. For some, having a sexual partner grab their hair in the throes of passion can be a turn-on, but if that happened while relaxing on the couch, it might get quite a different response.

Likewise, the reasons why we engage in sex aren't always about touch. Sex is important for reproduction. It can reduce stress and anxiety, contribute to feelings of escapism and fun, and improve our self-esteem. We can seek any mixture of these goals—and others that I've not listed—for ourselves or our partners at different times. Having a variety of reasons for engaging in sex is often viewed as part of having a healthy sex life.[13]

Whatever our reasons for engaging in sexual activity, our sense of touch is intimately connected to how we experience sex. Activities like kissing and sensual touching can be part of foreplay. The act of sex itself (either alone or with another person) also involves touch to intimate body parts—and for some people, nonintimate regions too. Sure, sex is more than simply an act of touching. But touch is an integral part of the process.

To understand the role of touch in sex, maybe let's start by asking how touch can lead to sex.

In 2012, researchers from the University of Alaska Fairbanks and Hobart and William Smith Colleges published a study called "Prelude to a Coitus: Sexual Initiation Cues Among Heterosexual Married Couples." The research involved asking couples to keep a sex diary for two weeks, detailing sexual activity and what they or their partner did to initiate it.

Physical touch was the most-used cue to initiate sex, with 77 percent of the research participants reporting that they engaged in touch to start sex over the two weeks. The types of behaviors that people engaged in included a range of examples. Some spoke of "holding hands." Others of "rubbing my breasts." Some were even more direct, indicating that sex started by "pressing up against him" or "prodding penis between legs."

Verbal cues like "asking to have sex" were used by 70 percent of the research participants, while nudity and undressing their partner were used by less than 20 percent of participants.

Touch was the most common signal to initiate sex for both men and women. And partners easily understood it, with 92 percent saying they recognized the intent of a tactile signal to initiate sex, and 91 percent saying it was a cue their partner often used.[14]

Why is touch so important to sex initiation? There are many reasons. From a sensory neuroscience perspective, intimate body parts like our genitals are sensitive to touch. The skin in areas of the body like the clitoris or the tip of the penis is full of nerve endings that respond to things like heat, cold, pain, and vibration: There's a reason why sex toys buzz!

The signals from touch receptors in our intimate body parts during sexual activities—both sex and foreplay—are sent via specific sensory nerves to the brain. These include the pudendal nerve, a sensory nerve thought to play a role in transferring information from the clitoris or the penis, and the hypogastric nerve, which travels directly to a part of the brain stem.

The brain stem is at the bottom of our brain. It looks a bit like a stalk and connects our brain to our spinal cord. It sends signals from the brain to the body and contributes to things like breathing and heart rate (two things that can change quite a bit during sex).

There are also nerves sending signals from other areas of the body that may be stimulated during sex, like the vaginal wall, the anus, and the scrotum. In women, the pelvic nerve plays a role in carrying touch signals from relevant body parts. In men, the pudendal nerve is important for this.

The key message here is that there are common and distinct nerves carrying touch signals from our genitals and nearby regions (called perigenital regions).

The number of nerve endings that respond to touch in these areas can vary from person to person. For example, someone could have fewer nerve endings in their genitals but more in other perigenital regions, like the butt. This has led to some suggestions that individual preferences for sexual touch may connect back to an individual's anatomy of sensory receptors in genital and perigenital areas, or to individual differences in cross-talk between nerves associated with touch in these regions.

Put a little differently, the reason some people might enjoy certain sexual activities more than others—like more butt-centric activities—could connect back to the physiology of tactile stimulation of these regions. Although, I must admit that I do not know of any study directly investigating these hypotheses.

What is clearer, however, is the brain basis of sexual touch. We noted already that the brain stem can be activated via the hypogastric nerve. There are other brain areas connected to sexual touch. Many of these have been identified by experiments examining sexual touch and orgasm while people have their brains scanned.

Conducting brain-scanning experiments on sexual touch is a little tricky. For those of you who have been in a brain scanner, you may be able to imagine why. You lie on a bed that often feels quite cold and plasticky and are slid into what looks like a giant doughnut ring, sometimes with something that appears to be a 3-D medieval mask on your face. Then there's a lot of noise, which some people have described as being like a jackhammer a few feet away. Not quite the description you'd hope to find in a book like *Fifty Shades of Grey*, unless you're into brain scanners—in which case, enjoy!

Despite these rather unfortunate parameters, scientists have conducted several studies on sexual touch and orgasm in brain scanners like these. These often involve self-touch. Armed with sex toys, research participants have been asked to go into the scanner and stimulate their genital regions, all in the name of science.

Much of the research on the neuroscience of sex and orgasm has been pioneered by Barry Komisaruk and Beverly Whipple, of Rutgers University in New Jersey.[15] Over the years, their research has shown that sexual touch and orgasm are related to extensive brain activity. As Komisaruk explained in an article published by the American news website Vox in 2015, when it comes to orgasm, "more than thirty major brain systems are activated. It's not a local, discrete event. There's no 'orgasm center.' It's everywhere."[16]

Patterns of brain activity during genital stimulation and orgasm can be quite similar in men and women. The somatosensory cortex is often involved in the buildup to orgasm, likely reflecting responses to genital touch.

But other brain regions are also important.

One brain area, called the hypothalamus, is thought to respond to arousal through sexual activities like touch. It contributes to the functioning of a range of hormones and neurotransmitters associated with arousal and positive responses to sexual stimulation. Some hormones that are thought to be released include dopamine, testosterone, and noradrenaline.

Changes in brain regions within the limbic system—like the amygdala and hippocampus—are thought to reduce fear and aggression before orgasm. At the same time, it is suggested that changes in prefrontal brain regions contribute to the deactivation of brain networks involved in self-evaluation, reasoning, and impulse control. All these support a reduction in fear and anxiety, which is often considered essential in leading up to orgasm.

The experience of orgasm itself has been linked with activity in regions of the brain associated with feelings of euphoria, while the release of oxytocin and dopamine contributes to the relaxing pleasure connected with experiencing orgasm.

The fact that orgasm can cause the release of hormones like oxytocin and dopamine speaks to another important message: Orgasms can offer benefits to our physical and mental health.[17] This is true whether the experience comes with a partner or simply by going solo. Some benefits connected to orgasm include better mood, better sleep, better skin, and better immune response. Put crudely, science tells us that orgasm can have many advantages: It gives us a nod to enjoy some consensual or solo fun; it's good for us, after all.

THE BENEFITS OF SEX FOR WELL-BEING

In all this talk about orgasms, we should also remember that the most important part of the sexual touch story is not necessarily the ending. The journey we go through by engaging in consensual sexual touch offers many benefits and opportunities for people to connect, de-stress, and escape the world around them. Sex gives people the opportunity to touch each other. To be close to one another. That need not involve an orgasm. The tactile journey through sex offers strong potential for bonding via touch. This can have valuable impacts on satisfaction and well-being throughout life.

A group where this has been shown quite clearly is in older adults.

It is often assumed that later life is less sexual. Yet many studies and surveys of older adults show that people remain sexually active for a long time. In her excellent book *Great Sex Starts at 50*, Tracy Cox provides an insightful and empowering discussion of how a robust sex life can contribute to health, well-being, and relationship satisfaction at different stages of adulthood.[18]

Some research resonates with many of Cox's conclusions. In 1997, a study of over 1,200 older American adults, with a mean age of 77 years, showed that 30 percent had been sexually active in the past month, and 67 percent were satisfied with their sexual life overall.[19] For our purposes, what is most important is that the sexually active and satisfied older adults showed higher levels of life satisfaction than older adults who were not sexually active.

More recent research shows that this relationship is not simply connected to having sex. Different types of tactile contact during sex can contribute to the relationship between sex and life satisfaction in older adults.

In 2019, researchers from Opole University found that while positive relationships exist between the frequency of sexual contact and overall life satisfaction in older adults, for many, it was simply the buildup that counted.[20] Engaging in activities like kissing and cuddling were preferred to having sex. These forms of tactile affection, which were linked to engaging in sexual activity but did not involve sexual intercourse, were just as good for global life satisfaction as engaging in sexual activities involving genital touch. By allowing people to touch, sex opened a way to impact well-being positively.

You might wonder if relationships between sex, affectionate touch, and well-being are only found in older adulthood. It turns out that affection facilitated through sex can benefit well-being throughout adulthood.

In 2017, scientists from the University of Fribourg conducted four experiments involving adults aged 17 to 64. They asked participants about the frequency of sex in their lives. Additionally, they asked about the amount of affection in their relationships.

They found a connection between frequent sexual activity and overall well-being: The more sex people had, the higher their well-being. Importantly, this was connected to the amount of affection they experienced in their relationships. People who had more sex showed more affection, and people who shared more affection were more likely to show benefits from sex to their well-being.[21]

To help make that a bit clearer, in two experiments, affection was measured by asking people how often they shared affective touch experiences with their partner, like hugging or kissing. The amount of affective touch that couples shared was critical in explaining the connections between sex and well-being. Engaging in sex more often benefited well-being more when the couples shared more affective touch in their relationship. Simply put, one way that sex promoted well-being was because it was a channel for affection to take place.

These findings help to show us that having sex can offer benefits to relationships and well-being. Yet these benefits are not all about the experience of sex itself. They can be about feelings of affection and affectionate behaviors shared through touch. Even if it is not possible to have sex, we can feel closer to our partners through forms of nonsexual intimacy like hand-holding, cuddling, and kissing. Tactile affection appears crucial in helping to unlock the potential benefits of sex to our well-being.

Make affection, not love

Many aspects of romantic relationships contribute to happiness, well-being, and satisfaction. Intimate touch between romantic partners and our satisfaction with how much we receive are critical parts of this process. It doesn't matter whether you are a young, middle-aged, or older adult: Tactile intimacy with a romantic partner helps us feel more satisfied with many aspects of life.

Our desires and needs for different types of touch are nuanced, though. When engaging in intimate partner touch, the idiosyncrasies of what you like versus what your partner likes should always be considered. It's important to check in with your partner about their feelings about intimate touch, particularly in long-term relationships where preferences can change as you grow older. Ask yourself: What's your favorite type of tactile intimacy? Has this always been the case, or has it changed over time?

Even with the people we are closest to, touch can be complicated. Partner touch is a behavior that people engage with readily but often don't pause to speak or think about. Given how crucial intimate touch can be to our happiness and relationships, we may need to encourage more open conversations about tactile intimacy to maximize the power of touch in our closest relationships.

To end this chapter, I've borrowed a phrase from Tracy Cox's *Great Sex Starts at 50*, discussing ways to engage in intimacy without necessarily having sex that involves intercourse: "Make affection, not love."

I want to expand upon this to consider a broader theme that we've seen throughout Part 2 of this book: how conveying affection and social support through touch can help mental and physical wellness. This raises an important question: How can we bring affection into the lives of those who desire more of it?

In *The Loneliness Cure*, Kory Floyd provides detailed strategies to try to overcome feelings of lack of affection in our lives. I've attempted to summarize and apply these to sharing tactile affection below.

1. **Understand**. Before we try to bring more affection into our lives, we must understand where we want to go. What is the affectionate touch we desire from different people? How do we approach affectionate behaviors towards them? To help with this, you may want to check out some of the further materials that I list in the Appendix (page 221).
2. **Communicate**. I suspect that reading that communication is important to relationships will not be news to you. But

if we want to share more affective touch with partners (romantic or otherwise), we need to communicate about it. It's good to share feelings, opinions, and expectations so that we can try to understand where each of our behaviors comes from. Try to remember that communicating is not just about what you are hearing; it is about taking time to make sense of and acting upon what people are saying.

3. **Notice**. Affection comes in different forms. If you identify that you and the people you have close relationships with differ in how you approach tactile affection, then sometimes the next step is to notice that affection may be shown in a different way. People may try to show us affection, but we don't interpret it that way. Not all forms of affection must be tactile: An affectionate note or someone making a nice meal can be signs of affection too. Noticing and acknowledging different forms of affection can make the diversity of affection people share become more meaningful.

4. **Invite**. If we want to bring a more affectionate touch into our relationships, we also have to try to invite it in ourselves. We can model being affectionate towards others, which can bring benefits: Some research suggests that engaging in affectionate behaviors can improve the mood of the person who initiates the affection and make them appear more trustworthy.[22] It is important that we emphasize the term "invite," rather than "demand." It is good to share your desire for affection with a close relationship partner. However, remember that people can differ considerably in how they approach affectionate touch, and increasing tactile contact can sometimes feel uncomfortable. Try to be patient, open-minded, and compassionate to your partner's needs.

PART 3

TOUCH TRAITS

CHAPTER 6

Touchy-Feely or Avoid at All Costs: Our Touch Personas

We keep returning to a recurring theme throughout this book: Touch is nuanced. We all have different desires for and experiences of touch. We differ in how much touch we give to others. We differ in how much touch we have experienced throughout our lives.

Before we go on, just pause for a moment and ask yourself how much you think the statements below describe you most of the time. Tick the relevant box next to each statement, from "Strongly agree" to "Strongly disagree." Some statements ask you to think about certain people in your life, like a partner, but if you do not have a partner right now, try to imagine how you might feel if you did. Be honest; there are no right or wrong answers.

		Strongly agree	Somewhat agree	Somewhat disagree	Strongly disagree
1	In an intimate relationship, I like to gently caress someone I care about				
2	If I know someone intimately, I enjoy the sensation of my skin against theirs				
3	If someone I don't know well puts a hand on my arm in a friendly manner, I feel comfortable				

		Strongly agree	Somewhat agree	Somewhat disagree	Strongly disagree
4	If someone I don't know well gives me a hug, then I'm okay with that				
5	Most days I get a friendly hug or a kiss				
6	When I have a romantic partner, I regularly share romantic kisses with them				
7	When I am out with friends, I like to touch them				
8	I greet my friends and family by giving them a hug				

How did you respond? Put simply, the more "Strongly agree" responses you give, the more favorable your overall attitudes and experiences to touch will be.

You may also notice that the quiz has a few themes within it. Numbers 1 and 2 ask about feelings towards intimate touch; 3 and 4 are about unfamiliar touch; 5 and 6 ask about engagement in affective touch behaviors; 7 and 8 ask about engagement with friends and family touch. You might score high on some items but not others. For instance, let's say you are someone who likes touch with friends and family but dislikes touch from people you are not familiar with. In this case, you might have given more agree-with responses for 7 and 8 but more disagree-with responses for 3 and 4.

Were you surprised at all by your answers? If I asked your partner or friend to fill out the answers to describe you, how do you think they would evaluate the questions?

Sometimes when people do quizzes like this, they realize that they don't often stop to think about touch in their lives. They may have certain thoughts about touch, but their behaviors may not always align with these.

This quiz is adapted from a measure we used in the Touch Test. It was originally developed from a full questionnaire on attitudes towards and experiences of touch by researchers based at Liverpool John Moores University.[1] In research studies, the original questionnaire is commonly used to understand individual differences in our attitudes and experiences towards touch. I've adapted it here to help us stop and reflect on perspectives and experiences of touch in daily life.*

We should keep in mind that our level of agreement with descriptions about how we find touch most of the time will rarely capture the richness of tactile experiences, which can change even when

---

* It is important to understand that the quiz is not a diagnostic, nor should it be used as a psychometric measure, as it has not been scientifically validated. Rather, I hope you will use this quiz to help you think about your own touch preferences.

it comes from a similar group of peers, such as our friends. As one person who experiences high sensory sensitivity explained to me during the Touch Test: "In most circumstances, I find touch overwhelming and powerful. With my partner, it is pleasant, sometimes deliriously so. I can enjoy a firm hug with friends, but pretty much anything else is uncomfortable or distressing. Certain touch sensations make me shudder to think about. One of the worst is attempting to dry a wet wooden spoon with a tea towel. Another is the fuzzy skin on a peach."

Nevertheless, to help understand differences in how we all think and feel about touch, it can often be helpful to begin by trying to reflect on our own thoughts and experiences (like you just did in the quiz above). After all, the better you know yourself, the more likely it is that you will be prepared for understanding others.

THE TOUCH PERSONAS TASK

To further consider how people vary in their responses to touch, it can also sometimes be helpful to think of the diverse touch personas we might encounter daily. To give you an idea, below are a series of descriptions of how different people might approach touch in their lives.[2] Ask yourself, which persona best represents you?

> **Charlie** can't help but touch. A squeeze of another's arm during conversations, a hug, a kiss on the cheek to say goodbye. Charlie is straight in there touching others, be they friends, partners, family members, or strangers. They seem to love touch and feel comfortable touching or being touched by just about anyone in social settings.
> **Kei**'s touch persona is best described as cautious and considered. Sure, they find touch pleasant, but they only really like to be touched by family and close friends. Even then, there are clear boundaries, and they touch with care. The idea of

being touched in a friendly manner by someone they do not know makes them uncomfortable.

Taylor feels uncomfortable when touched by other people. It doesn't matter whether this is a friend, family member, or stranger. They don't understand how people can enjoy the sensation of someone stroking their arm or hugging them. Even more so, why would anyone voluntarily let a stranger touch them during a massage?

Zayda wants more touch in life but feels uneasy when engaging in touch with other people. They are nervous and tend to hold back from touching others, despite their desires. For instance, they would like to hug their partner goodbye before they leave for a business trip, but it's not something they do, and Zayda doesn't feel confident to start.

Drew is naturally reserved when it comes to touch, but their persona is adaptable based on the setting. For instance, most of the time, Drew would wince at having to kiss someone they do not know well on the cheek as a greeting. Still, Drew travels a lot for work to regions of the world where this type of behavior is more common. In those settings, they feel comfortable with kissing as a greeting because they've adapted to thinking of it as routine, like handshaking before a business meeting back home.

Which persona did you choose: Charlie, Kei, Taylor, Zayda, or Drew? Do you know anyone in your life who you would describe as matching one of the personas you did not choose?

These personas are not just about people we know. We can use them to consider how we might interact with different touch personas in real life. Consider how you would feel if you went on a date with someone like Taylor, who rarely touches you. Or maybe if you had to approach a situation involving someone like Charlie, who you can see coming in for a hug you don't want from miles away.

When I tell people that I'm a touch scientist, one of the most common questions I'm asked is: Does that mean you can tell me how I can respond to a hugger? I wish I could say that science has the answer. Sadly, few scientific studies have explored how to approach situations in which people with different touch personas interact.

Instead, the broader science of interpersonal communication has taught us that one of the best things we can do to support our social interactions with others is to communicate carefully and clearly about our preferences.

Take Zayda's conundrum of desiring hugs from their partner but not knowing how to start. The simple suggestion could be for us to encourage them to try initiating a hug one morning—after all, we know that expressing our feelings of affection can be good for stress and mood. But the confidence to engage in that first hug might be a challenge. What alternative suggestions could we offer? Perhaps we could encourage Zayda to communicate with their partner about their touch preferences. If that's a struggle, we could start by giving them space to talk openly with us about touch.

Sharing thoughts about our touch preferences doesn't always have to be verbal. Imagine if the tactile Charlie were moving around the room at a social event, air-kissing guests on the cheek. You know this level of tactile closeness is not for you. The time to verbally express this is short. What can you do? Social etiquette experts suggest that one good option is to keep your hands in your pockets and smile, or give a solid but friendly gesture of pause with your hand. Personally, I'm also fond of a lean or step back with a nod of acknowledgment.

THE LEAGUE OF EVIL EXES

While we might be waiting on science for the answers to the best way to deal with an unwanted hug, studies in psychology have given us considerable insights into reasons why our touch preferences can vary. Researchers have asked what psychological traits are more likely to be found in someone who feels positively about touch than in someone who does not.

One of the most striking factors is how people approach attachment in their relationships. To help me explain this, and to avoid me getting too personal about people in my life history, let me draw on one more movie: *Scott Pilgrim vs. the World*, Edgar Wright's cult classic film from 2010, based on the graphic novel by Bryan Lee O'Malley. The film sees Scott meet the girl of his dreams, Ramona Flowers. The trouble is, Ramona comes with some baggage—seven evil ex-partners whom Scott must defeat "in battle" to date her.

There are many reasons to love this romantic action-comedy movie. My reason for mentioning it here is simply that it's full of a wonderful variety of different attachment styles. Some of Ramona's ex-boyfriends are archetypal avoidant characters. They want space in their relationships and appear ready to bolt if someone gets too close to them. Others are what we might colloquially call clingy. They look for reassurance and follow Ramona around. Displaying behavior consistent with having difficulty trusting others in relationships and worrying about being abandoned are hallmarks of people with an anxious attachment style.

You might recall that we came across anxious and avoidant attachment styles when discussing touch hunger in Chapter 4. People with an anxious attachment style may be more likely to enjoy affective touch. This can mean that unmet touch needs can be more detrimental for people with higher anxious attachment styles than those with avoidant attachment styles. For instance, during the social restrictions of the COVID-19 pandemic, people with a higher anxious attachment style were found to express a greater need for touch than people with higher avoidant attachment.

In contrast, people with a more avoidant attachment style did not feel the loss of touch during COVID-19 pandemic restrictions as much. This makes some sense because people higher in avoidant attachment tend to fear intimacy and prefer to maintain independence from their partner. They may dislike affective touch and can have a lower desire for touch in relationships.

In short, affective touch matters much more to some individuals than others; therefore, having this taken away has different effects depending on needs and preferences.

People with anxious attachment also differ from those with avoidant attachment styles in how touch impacts broader behaviors, such as pain reduction after receiving pleasant touch. One study led by researchers based at King's College London and University College London saw people's brain responses and self-reported experience of pain measured while they experienced a painful sensation induced by a laser. Yes, a laser!

This method, called a laser-evoked-potential study, involves zapping people with infrared heat pulses. While this is happening, participants are asked to self-report their pain levels, and brain signatures connected to pain processing are measured.

The researchers found that people higher in attachment anxiety showed more benefits from pleasant touch in response to heat pain. In contrast, people with higher attachment avoidance showed a lower reduction in brain responses to heat pain when receiving pleasant touch.[3]

This takes me back to Chapter 3. You might recall that some forms of touch can make people more at ease or reduce painful experiences in specific situations (e.g., undergoing medical procedures). We might expect that some of the findings we encountered showing that touch can impact health could vary according to the attachment style of the person being touched.

*

The importance of attachment style also goes a step further. It can help explain other social dynamics, such as how we perceive the quality of touch in our relationships.[4]

In 2020, researchers from Binghamton University studied the impact of individual differences in attachment style and touch satisfaction on relationship quality in married couples. In other words, does the interaction between tactile affection and attachment contribute to happier marriages?

In the research, 184 married couples from New York were asked to complete surveys about their relationship satisfaction, attachment style, touch satisfaction, and engagement in routine nonsexual tactile affection with their partners.

The researchers found that greater touch satisfaction was related to more marital satisfaction. That is to say, people who were happier with the amount of affectionate touch in their relationships were also more satisfied with their relationships overall.

Taking a deeper look at the results revealed a more nuanced message. The overall relationship between tactile affection and touch satisfaction seen in these New York couples was heavily influenced by the attachment style of the husband in the relationship.

Yes, on average people who were happier with the amount of affectionate touch in their relationships were also more satisfied with their relationships overall. But husbands who had more attachment anxiety were more likely to be less satisfied with touch when engagement in tactile affection was low. In contrast, if the husband in the couple experienced high attachment avoidance, then touch satisfaction was less likely to impact feelings of marital satisfaction.

This makes sense when we consider that people with greater attachment anxiety tend to be more likely to enjoy affective touch. If you are or have a husband who enjoys a cuddle, you may find that the same husband will be less satisfied when this form of affection is absent.

These results help to remind us that the sweet spot for relationship satisfaction and touch is somewhat nuanced. It is not a simple case of more is better or less is worse. Instead, it is about matching the quality of touch each person receives to their desires.

This is a conclusion that has been supported by other research. In 2020, researchers from the City University of New York asked romantically partnered undergraduate students to keep a daily diary of the touch behaviors they exchanged with their partners over 10 days. They were also asked about their attachment styles and several measures of relationship well-being, like how close they felt to their partners.

Receiving partner touch benefited relationship well-being. But attachment styles were found to influence the extent to which participants benefited from touch behaviors. The gains in relationship well-being from touch were strongest in people high in anxious attachment. Essentially, tactile affection might benefit couples, but those benefits can be amplified in people with higher attachment anxiety.

Affectionate touch may also help romantic couples view activities they engage in more positively. But again, this may be affected by attachment styles. In 2022, researchers from Syracuse University invited couples into a living room laboratory: a lab that looks like a living room.

While hanging out in the living room lab, the couples engaged in a conversation about activities that people don't always make time for. Crucially, half of the participants were encouraged to sit close and touch each other affectionately during this conversation. The other half were not. Instead, they had to take notes on what the speaker was saying: They could not touch, but they did have to listen carefully.

After the initial conversation, the couples had 10 minutes of free time together. They could interact as they saw fit while they waited for a new experimenter who was running late to arrive. Little did they know that this free time was an all-important ruse on the part of the researchers. The researchers wanted to know whether prior engagement in touch would impact how the couples felt they spent this time together.

When asked later about the 10-minute free time, the couples who had touched beforehand reported that they spent the time engaging in positive activities together. They were more likely to agree with questionnaire statements like "My partner and I did something fun together."

What struck me in this experiment was that when an independent observer watched videos of the couples interacting during this free time, they could not notice any differences in the types of activities they engaged in. Essentially, couples engaged in similar mundane activities during their free time, but those who touched before found the interactions more fun and intimate.

What's more, when some of the couples were tested again a week later, the touch group still indicated having more positive interactions with their partners, an incredible demonstration of the power of touch on relationship perception.

But there was a twist in the tale. The positive perceptions shown by people in the touch group were influenced by attachment style. The benefits of positive perceptions of partner interactions associated with touch were found most strongly for people with greater attachment security. Essentially, the likelihood that people may perceive everyday interactions with partners positively was impacted by interactions between affective touch and attachment styles.

Collectively, these findings point to a strong relationship between touch satisfaction, relationship satisfaction, and individual differences in our desires for touch that may emerge due to how we approach attachment in our relationships. They remind us that we should not assume touch will benefit everyone equally.

This brings us back to a message from our chapter on tactile intimacy: Couples might find it helpful to take the time to establish what forms of touch they enjoy together. By communicating and learning about these desires, they could make informed choices about forms of tactile affection that support both partners' needs. This could help with feelings of touch satisfaction, which in turn may impact feelings of relationship satisfaction.

> **The paradox of affection and avoidant attachment**
>
> From the research on attachment style and touch in romantic couples, we might erroneously conclude that people with more avoidant attachment styles do not benefit from affectionate behaviors. To be clear, this is not correct. Evidence indicates that although people with an avoidant attachment style may hold less favorable views about closeness, they still benefit from expressions of support from their partners. For instance, in 2021, researchers from the University of Lausanne examined how tactile experiences in romantic relationships contribute to well-being. They found that attachment avoidance was related to less frequent touch in romantic relationships. Despite this, there were still positive relationships between engaging in touch and well-being in individuals with high levels of attachment avoidance. In other words, the relationship between touch and psychological well-being was positive, regardless of levels of attachment avoidance.[5] People with attachment avoidance may sometimes show a lower desire for touch, but the limited amounts they receive can be beneficial. This helps to remind us that it is not necessarily the frequency of touch that matters most for the beneficial effects to occur. More is not always better. Instead, it is helpful to ensure people get the quality of touch that aligns with their own desires.

THE BOLD, THE TRUSTWORTHY, AND THE CURIOUS

What would you say if I asked you to describe your personality? Are you talkative? Enthusiastic? Curious? Anxious? You'd likely agree more with some of these words than others. And just like with touch, you might also point out that it depends a bit on the context that you find yourself in—you might be a talkative person at home but change this behavior if you are somewhere that calls for you to be quiet, like in a library.

Personality traits reflect the characteristic patterns of a person's thoughts, feelings, and behaviors. They have been the subject of psychological research for many years.

Early personality researchers in the 1930s explored more than 4,000 different personality traits.[6] Years of research and some clever data analysis techniques* have helped reduce this number. These days, psychologists tend to agree on a smaller set of broad dimensions of personality.[7]

You might have heard of the Big Five personality traits. These are suggested groupings for different personality traits that have been identified around the world. They are agreeableness, conscientiousness, extroversion, openness to experience, and neuroticism.

Each trait is thought to capture a range of characteristics that tend to cluster together. If you are highly agreeable, you might be the type of person who is trustworthy, willing to compromise, and helpful. You, and your conscientious friend, might be focused, have high self-discipline, and always like to be prepared.

Someone who is high in extroversion is enthusiastic, full of energy, and chatty, while openness to experience describes

---

* These techniques allow researchers to reduce thousands of variables into smaller groups of related personality traits (e.g., pairing traits like talkative and energetic into a shared category).

the type of person who likes to try new things, who is curious, receptive to emotion, and appreciates adventure. Last but not least, a person with high neuroticism tends to experience negative emotions, worries a lot, and has a low tolerance for stressful situations.

The Big Five personality traits are pretty stable throughout the life span. They have also been seen in a variety of groups of people worldwide. Some research suggests that a person's levels of extroversion, openness to experience, and neuroticism can show stability across 45 years.[8]

Our personality traits have also been linked to various behaviors in everyday life, including our social relationships. Given the importance of touch in social interaction, it would not be surprising to observe differences in how people evaluate and engage with touch because of their personality traits. Indeed, this is something that scientists have studied.

In 2008, Sam Dorros and researchers based at the University of Arizona conducted a study investigating how the Big Five personality traits contribute to perceptions of touch from a romantic partner. They also asked what was more important in predicting positive perceptions of touch—the personality traits or the gender of people in the research.

Just over 300 adults completed questionnaires that measured their personality traits. They were also given a picture of a body and asked to indicate how experiencing touch from their partner on different body parts would make them feel. For instance, does it feel pleasant to be touched on the shoulder, the stomach, or the neck?

The results showed that specific personality traits were linked with different perceptions of touch. People with higher agreeableness showed more positive perceptions of touch to any region of the body, intimate or not, while differences in the personality trait of openness to experience predicted how positively people rated touch to non-intimate body regions.*

On top of this, the researchers found that personality differences were a stronger predictor of how people felt about experiencing touch from their partner than their gender or their current level of relationship satisfaction.[9] This fascinating finding shows how influential psychological factors like our personality traits can be in how we might approach touch in romantic relationships.

One of the things that surprised me about the results from Sam Dorros's study was that some of the personality traits that might be expected to play a role in perceptions of touch did not do so. Intuitively, you might think that more outgoing people perceive touch differently from those who are less extroverted.

Some research would back this prediction up. People who score higher in extroversion tend to engage more in social contact. They also show differences in how their brain responds to touch: for instance, responses in the somatosensory cortex that have been connected to differences in tactile sensitivity.[10] Yet in the Arizona study, extroversion didn't appear to predict touch preferences from a romantic partner.

I was stunned. I wondered if part of the reason for this may have been because the study focused solely on partner touch. What would happen if people were asked about their attitudes and experiences towards everyday social touch? For instance, touch from strangers, friends, family members, and professionals. My curiosity was sparked.

Luckily for me, the Touch Test allowed me and my team to look at this question directly. We found that extroversion is a

---

* In this study, the authors used a form of data analysis that enabled them to cluster different body regions together based on participants' responses. Intimate regions included the chest, stomach, groin, thigh, and butt.

very powerful personality trait in explaining how individuals view everyday social touch.

Extroversion emerged in our study as the strongest personality trait predictor of attitudes towards day-to-day social touch, such as shaking hands. People who reported being more extroverted reported more positive attitudes to and experiences of all types of touch, especially public forms involving friends or strangers.

In the same data set, more extroverted people reported having more positive thoughts on engaging with tactile treatments to support medical health and self-care,* such as massage and beauty treatments.[11]

So, what all this suggests is that our personality traits influence how we think and feel about touch. But the relationship is not an all-or-nothing scenario. Certain personality traits matter more than others, depending on the situation. You'll recall the importance of agreeableness and openness to new experiences in cases involving touch from a partner in the Arizona study. In other scenarios, like social touch involving a range of different people, other personality traits may be more important: for instance, the importance of extroversion that was shown consistently in the Touch Test.

Put a little differently, touch is far from straightforward. Its nuanced nature means that we must be mindful of the delicate balance between person and context when tailoring our tactile exchanges from one person to the next.

THE PEOPLE WHO SHARE YOUR PAIN

An underlying message of this chapter has been that our likes and dislikes around touch are nuanced and vary from person to

---

* Other relevant differences include age and gender.

person. We should always take other people's personal preferences into account.

One topic we've not touched on so far is how types of neurodivergence may impact how people engage with touch. People who are neurodivergent are considered to differ from the dominant societal view of what might be regarded as the norm in the way the brain engages in mental processes.

Since Judy Singer's early use of the term "neurodiversity" in the 1990s, society has become increasingly aware of the importance of neurodiversity and the potential benefits of embracing people who may think and experience life differently from the status quo.[12] Although we may not commonly think of neurodivergence in a similar way to things like differences in personality types, several examples have been linked to differences in how people experience touch.

One example is that many autistic people experience touch differently from non-autistic people.[13] Sometimes this can be shown as an increased sensitivity. For instance, people who are oversensitive to touch may struggle with the feel of certain fabrics.

There can also sometimes be an under-responsiveness to touch. For these people, touch may not register, or they may show difficulty attending to tactile experiences.

Tactile sensitivity can also vary according to the type of touch and the context in which it occurs. Some studies report that people on the autism spectrum can rate textures as more pleasant while being more sensitive to vibration and painful touch. This finding also aligns with some of the experiences we've read from non-autistic people throughout this book.

These different profiles of tactile sensitivity can contribute to changes in behavioral responses. An undersensitive person might engage in sensory-seeking behavior like frequently touching objects. In contrast, a person with increased sensitivity might experience a sensory overload. This could contribute to behaviors that may see them withdraw from situations where they feel overwhelmed by touch.

Given these factors, some researchers have highlighted that although certain forms of touch, like hugging, may offer benefits for health and well-being for some of us, we should not assume that they will benefit everyone. People with high or low sensitivity to touch may find interpersonal tactile experiences too intrusive.

That's not to say that we shouldn't engage in touch with people who have different sensitivities to touch. The two most common words that autistic people used to describe what touch meant to them in the Touch Test were "comfort" and "discomfort," two polarized responses that act as a window into a much bigger picture about touch and autism.[14] The experiences of autistic people can be highly variable from person to person. We should not assume that within neurodivergent groups, everyone has a similar experience. Instead, we need to ask what is best for the individual—what are their personal needs and preferences?

This message was illuminated for me in a conversation I had with Jenny, a social worker. Jenny's role involved working closely with vulnerable children, and much of her caseload involved children with high sensory needs.

Jenny told me, "From a professional perspective, I must be mindful of how I use touch to communicate. I work with children from birth to eighteen years old—most have complex and very high sensory needs. What I am always mindful of is what are the ways that the child I am working with communicates. Connecting with them is about how I adapt who I am and how I approach them and communicate in a way that fits in with them rather than expecting them to fit in with me. Many of the children I work with have high sensory needs in terms of the need to touch things to seek comfort or to self-regulate. So, I tend to use tools and materials as a way to touch: stuff like slime or water, any kind of materials that I know will work for the young person that I am interacting with. I let them take the lead, exploring the textures and playing with them. I always research the child's best form of communication and what kind of sensory materials I should take with me. I always let them take the lead. I find that using the senses like this breaks down a barrier and builds a rapport—and relationship building starts from there."

Another neurodivergent group displaying differences in how they process touch are people who experience synesthesia. These people experience an unusual blending of the senses.[15] They might see colors when they listen to music. They could perceive colored halos around people when they show different emotions. They might even experience taste when they hear words.

Synesthetic experiences are present from early childhood and can have remarkable consistency throughout life. As one synesthete, James, explains: "I've tasted sound, and word sounds, for as long as I can remember. So, the experience feels as natural and as normal as breathing or being able to see, touch, smell, or hear. The only thing I'd consider unusual is that the experience is a real mouthfeel, not just a simple association. It feels like I have something in my mouth when obviously I haven't."

James experiences word-taste synesthesia—one of at least 80 types of synesthesia that have been described. "My very first memories of tasting words go back to when I was around four years old," he continues. "I used to travel to school daily on the London Underground system with my mother, and I have very vivid memories of tasting rhubarb whenever I heard the sound of a Tube train pulling out of a station. I recall the daily journey consisted of getting on the train at a station that tasted of Dolly Mixture sweets. After a short ride, getting off at another station that tasted of raw sliced potato."

James's synesthesia involves sound evoking taste. There are many more combinations that we know of. One of the more common types is vision evoking feelings of touch. Some people report that when they see someone else being touched, they experience first-hand sensations of touch on their own body. If these people see a couple holding hands, they would feel a corresponding sense of touch on their own hand.

This experience, known as mirror-touch synesthesia, affects around 1.5 percent of people. That's over one million people in the UK. A remarkable number, especially because you would not even know it.

People who experience synesthesia are often surprised to hear that their experiences differ from those of others. Synesthesia is reported to be present from childhood, so why would you know that other people don't taste words or feel touch when they see it happen to other people unless someone told you that it is unusual.

In his book *Mirror Touch*, Joel Salinas provides a moving and thoughtful memoir of his life as a medical doctor who experiences mirror-touch synesthesia.[16] He describes his experience of touch through sight as feeling a sensation of touch that is experienced on his body in response to seeing someone else being touched, as though looking in a mirror. This sensation is often most strongly felt when face-to-face with the person or thing being touched.

Salinas's account is like accounts I've heard from many people who experience mirror-touch synesthesia. You might imagine that these types of experiences need careful managing. As another mirror-touch synesthete, Mary, explained to me in an article for *Aeon* magazine: "I hate it when my husband watches violent movies. I cannot watch them because I feel overloaded. This is obviously not a pleasant experience, and it's a downside to my synesthesia. The upside is that I also experience the nice touches, the caresses and the hugs. None of the experiences lasts for long, and for that, I am grateful."[17]

I've been lucky to work for over a decade with people who have mirror-touch synesthesia. My research collaborators and I have conducted some of the first work showing behavioral and neurological differences connected to this unusual interpersonal sharing of touch, like more emotional empathy and changes in body representation.

We've demonstrated that mirror-touch synesthesia relies upon the mechanisms that we all use when we observe touch to other people: the vicarious touch brain networks that we read about in our discussion about touch hunger.

You might remember that when we watch other people experiencing touch, we tend to recruit similar parts of our brain as when we experience touch ourselves. What appears to happen when someone with mirror-sensory synesthesia watches other people being touched is that they overactivate this vicarious touch system, to the point where they feel tactile sensations themselves. This has led some to suggest that the vicarious tactile experience shown in mirror-touch synesthesia is an extreme endpoint of a continuum in how we share the states of others.[18] It's an endpoint that can lead to an unusual and powerful sharing of tactile experiences.

There's one final experience I'd like to talk about where feelings of touch can be triggered by another sense. That experience is called autonomous sensory meridian response, more commonly known as ASMR. Although it is not strictly classed as a form of neurodivergence, the experiences connected to it are a fascinating window into individual differences in tactile experience.

In ASMR, tingling sensations are triggered by auditory (e.g., whispering), visual (e.g., gentle hand movements), and tactile experiences (e.g., tracing fingers on the body).[19] Some of my favorite ASMR triggers are videos from American artist Bob Ross. Listening to the relaxing tone of his voice, mixed with gentle brushstrokes, can in some people evoke tingling sensations around the back of the head and neck that radiate across the body. If you find yourself triggered to experience ASMR, you might also get a feeling of relaxation or euphoria from the sensations.

In recent years, ASMR has become something of an internet sensation, growing from a relatively unknown experience decades ago to an online movement involving millions of people: Some online ASMR videos on sites like YouTube have trillions of views.

A creative online community has emerged. This community is served by ASMRtists who intentionally develop content that can trigger the calming, tingling, sometimes euphoric sensations from sound and vision.

Alongside this ASMR community are scientists like me who have been lucky enough to study ASMR for the first time. In the space of a few years, scientists have demonstrated that when people experience ASMR, they show physiological and behavioral activity patterns that are consistent with their self-reported experience of tingles and complex emotional responses. These include heart rate reductions and pupil diameter changes (a physiological correlate of norepinephrine release that is hard to fake).

ASMR experiences have also been related to changes in brain activity connected to attention, emotion, and sensory experiences. In other words, the tingles of ASMR appear to map onto distinct physiological responses. They show a reliable and physiologically rooted experience.

Beyond this, the experience of ASMR has also been connected to self-care. Anecdotally, many people are reporting that they use the sensations induced during ASMR to de-stress and reduce anxiety, or as a substitute for lack of contact with others in their lives. In one study published in 2015, researchers from the University of Swansea found that 70 percent of ASMR responders said they watch ASMR videos to deal with stress, and 98 percent use them to relax. As such, labs worldwide are starting to examine the potential of ASMR as a tool for self-care.

There are few published studies in this area to date, but the existing studies point to the importance of more research. For instance, in 2022, researchers from Northumbria University examined the impact of watching ASMR videos on state anxiety in a group of 64 participants. State anxiety was measured using a questionnaire asking people how much anxiety they felt at that moment. The questionnaire was administered before and after participants watched a five-minute ASMR video. The participants were also asked whether they experienced ASMR during the video.

The researchers compared state anxiety in those who experienced ASMR tingles and those who did not. They found a reduction in state anxiety in people who experienced ASMR. In other words, experiencing ASMR helped with anxiety.

These findings are not without limitations: Recruiting people from online ASMR forums means a risk of the sample not being representative of results in people unaware of ASMR. It's possible that if you did not know about ASMR, you might find experiencing the sensations or watching the videos strange. There was also a lack of placebo control or people watching non-ASMR videos as an intervention, limiting the inferences we can draw. Nevertheless, the results do point to what is likely to be an increasing trend for research to explore the potential of ASMR as a source for self-care.

Making the exception the norm

When we hear about scientific findings, a headline statement often grabs our attention. It's easy to take these snapshot headlines as the truth behind a study, but the headline can only ever be a summary statement reflecting what happened to most people within a given study. If we're lucky, the report below the headline might provide more depth into the reported study. Yet in the fast world of rapid news and character limits, it can be challenging for any report to give us the whole picture. And even in a long-form piece of writing like this book, there is a necessary amount of summarizing and generalization, however hard I might try to contextualize the data and its limitations. Sometimes a summary is the best way to translate information to a broad audience.

What this means is that when we engage with any reported study result, it is crucial to stop and ask ourselves questions about how far we can generalize the inferences we make. This chapter has shown us just how critical individual differences can be in the case of touch. For some, touch is their favorite sense; for others, their very worst. All sorts of factors may subtly change how touch is perceived and interpreted. These include, but are not limited to, attachment style, age, gender, neurodiversity, personality, and previous life experience. These individual differences could change the outcomes of touch with another person and even influence how we experience watching touch between others.

Naturally, this makes touch complicated. But it also opens our eyes to how our differences make the world much more vibrant. We can widen our perspective by being open and learning about the nuances that contribute to how we each approach touch. We can actively connect with the world around us in a more engaging way. A good starting point to help us do this is to learn about our own touch preferences. To help with this, you may want to draw on the quiz or the persona task found at the start of this chapter, or some of the further materials offered in the Appendix.

CHAPTER 7

Touch Culture: How Our Backgrounds Affect How We Perceive Touch

I'm a coffee lover. Not quite a coffee snob, but I'm such a regular in my local coffee shop that I have recently started going to the gym with my favorite barista.

Today, sitting inside the shop on a rainy morning, I'm enjoying the feeling of the sunshine-yellow coffee cup in my hands. In the inclement UK weather, one of my favorite parts of winter is days like these. Cold outside but warm inside. The silky smoothness of my oat flat white mixed with the coffee's slightly bitter dark chocolate notes. All combined with the warming sensation of my fingertips surrounding the cup. Bliss!

At this moment, I'm not just enjoying my coffee. I'm also engaging in a bit of people watching. I'm looking around the shop, conducting a small experiment. I'm watching to see how often people touch one another.

Before you judge me for being a bit creepy, let me explain why. In 1966, a well-known psychologist from the University of Florida called Sidney Jourard conducted seminal investigations into cultural differences in social touch.[1] One part of his work involved observing pairs of people in coffee shops in Gainesville, Florida; London; Paris; and San Juan, Puerto Rico. The findings showed that people from some countries would touch each other more than others. For instance, in San Juan, couples touched each other around 180 times per hour, but in London, this average was closer to zero. A rather dramatic difference.

Even now, I am still amazed by Jourard's report. Sure, I was not alive in 1966, but to touch 180 times per hour would mean roughly three touches a minute. I've never been to San Juan, but that seems like a lot!

The idea that people in a London coffee shop might not have touched at all is also surprising. This and the report that people in Gainesville touched twice an hour seem at odds with my experience of touch in coffee shops in the UK and America.

Naturally, many things have changed in the world since the 1960s. For what it's worth, my own people-watching experiment counted 10 touches between one couple in 10 minutes. I figured it best to stop watching after that; back to the flat white.

One of the interesting things about Jourard's study is that it is often used as evidence of cultural differences in touch. Yet to my knowledge, Jourard never actually made strong claims that this was the case. In the original publication, he only briefly mentioned these results.

A more recent study conducted in 2017 paints a slightly different picture. This research examined the touching behavior of mixed-gender romantic couples in rural towns and urban cities in America. The researchers visited coffee shops in three rural towns and three cities in four states: Georgia, Alabama, Tennessee, and Mississippi. They watched three random couples for at least 20 minutes each, recording the number of touches, the site of touch, and how long couples touched. Not too dissimilar to what I just did, but with much more scientific rigor than my casual observations.

They found more than double the number of touches in city coffee shops than in rural towns. The time romantic couples spent touching in city coffee shops was also higher: typically, 25 seconds in rural locations compared with one minute of touching by couples in city coffee shops.[2]

This study's methods are slightly different from Jourard's original hourlong observation in the 1960s; they also point to a different data pattern. Averaging across all the romantic couples, whether in a city or a rural coffee shop, the results indicated just over three touches every twenty minutes. That's likely nine touches in an hour in American coffee shops today. It seems that it might be worthwhile to have a more formal replication of Jourard's original observations: If anyone would like to fly me to Puerto Rico to do this, I would not object.

WHERE IS IT APPROPRIATE TO TOUCH?

Whatever we think of Jourard's observation in the 1960s, I suspect we'd all likely agree with the view that people use touch in various relationships worldwide, from intentional tactile exchanges with loved ones to accidental or incidental touch with strangers on the daily commute. We often associate different countries with different approaches and attitudes towards touch in interpersonal settings.

We might think of some cultures as being more tactile than others. But how true is this in practice? Do our thoughts on the acceptability of touch from other people differ according to where we come from?

One of the leading authorities on touch across different world regions is Juulia Suvilehto from Linköping University. Suvilehto's research team is fascinated by the question of how humans establish and maintain social bonds.

In 2015, Suvilehto led a study with other scientists examining how people from different European countries felt about the appropriateness of touch.[3] They surveyed over 1,300 people from several European countries—Finland, France, Italy, Russia, and the UK—asking participants questions about where on the body they would allow different people to touch them.

Each person who took part in the research was shown a silhouette of a human body and asked to color in the regions of the body where they felt touch would be appropriate. They were asked to do this with four people in mind: a stranger, a friend, various relatives (mother, father, siblings), and their partners if they had one at the time.

Suvilehto's team asked about touch from different people because they wanted to see whether feelings about what was or was not acceptable changed depending on the relationship with the person doing the touching. The strength of these relationships was also measured by asking each person to indicate how close they felt to the person they were thinking about when answering where they thought it was appropriate for that person to touch.

To illustrate, imagine if I gave you a picture of a body right now and asked you to point to each part that you thought would be okay for a friend to touch. I'd then ask you how strong your emotional bond was with the friend you were thinking of.

Intuitively you might imagine that people would think it was more acceptable for a greater range of body parts to be touched by people they felt closer to. This was the case: Emotionally closer individuals were allowed to touch more body regions.

At the same time, touch from strangers was limited mainly to the hands and upper torso. To put it differently, we find touch from family members and partners that we are closer to more acceptable than touch from other people.

There were also some common taboo zones where touching was not allowed. I suspect you might be able to guess where these regions might have been: the genitals, the butt. But of course, these areas did vary according to the nature of the relationship with the person involved in touching. Genitals were a no-go zone for extended family, males in close family, acquaintances, and strangers. Touch to the butt was unacceptable for males in extended family, acquaintances, and strangers.

Importantly, geographical influences were minor. The body parts considered appropriate to touch were similar for adults from all the countries involved. People allowed for more touch in close relationships irrespective of where they came from. There were also similarities in the non-acceptable taboo zones.

In other words, although we might think there are large differences in the acceptability of touch across different countries, there were some striking similarities in the five countries that were tested.

When I first read about these findings, I wondered whether part of the reason for the similar responses in Finnish, French, Italian, Russian, and British adults was that they were closer in cultural connection than we might first think. Yes, there are differences between these nations, but they are all European countries commonly associated with Western cultural perspectives.

Would differences be more pronounced if countries from different continents were compared? For instance, historically, it has been found that Japanese adults engage in certain forms of social touch, like handshakes and hugs, less frequently than adults from some Western countries.[4] Are there differences in the allowance of touch between people living in predominantly Western compared to Eastern cultures?

Thankfully, in 2019, my curiosity was satisfied. Suvilehto's team published a follow-up experiment comparing UK and Japanese adults.[5] As with the original study, the people that took part were asked to color regions of the body where touch would be allowed from different people—strangers, friends, relatives, and partners. They also rated their emotional bonds with the people they were thinking about when answering the question.

The results showed that even in two countries separated by over 5,700 miles, the pattern of results as to where people felt touch was acceptable was similar.

Much like their original findings, the researchers found that touch was more permissible when people felt closer to the person doing the touching. This was true among both Japanese and UK adults. Women were also allowed to touch bodies more than men in both countries.

There were more subtle twists to the pattern of data, however. Yes, the main findings showing relationships between emotional bonds and where people were allowed to touch were similar between UK and Japanese adults. Yet when asked how pleasant they would find touch from each person (friend, stranger, relative, or partner), the overall rating was lower in Japanese compared to UK adults. Essentially, Japanese adults were less keen on touch, regardless of who was touching them.

This was further complicated by the gender of the person doing the touching. Both male and female adults from the UK preferred touch from women. In Japanese adults, women preferred touch from other women, while men did not.

The UK adults also indicated a greater allowance for female family members and friends to touch their faces than Japanese adults. This might be because UK adults also tend to touch their faces more than Japanese adults. It's therefore possible that faces are viewed, in general, as a more accessible region for touching in the UK.

These findings show us that relationships between culture, acceptability, and preferences for touch are not simple. There are similarities in how emotional bonds impact what we view as acceptable touch. There are consistencies in our taboo zones and the people we allow to touch them. Yet there are more subtle differences in perceptions of pleasantness and the acceptability of touch depending on where we come from.

The results from the classic coffee shop studies in 1966 through to the work of Suvilehto's team in 2019 show that although there may be subtle differences in how people engage in touch depending on where they come from, the reasons for this are complicated. They are connected to the type of touch people engage in, their relationship with the person they are touching or being touched by, the context, and various cultural and social values that may contribute to a person's identity.

Put a bit more simply, people are complex. It is hard to pin anyone down based solely on where they are from. We shouldn't pigeonhole people. Therefore, we shouldn't expect straightforward relationships between touch and culture.

CONVERSATIONS WITH FRIENDS

Hands, face, space. UK readers will be familiar with this slogan from a public information campaign launched in 2020 to urge the public to engage in behaviors to reduce coronavirus spread during the COVID-19 pandemic. People were encouraged to wash their hands, cover their faces, and make space to control infection rates to avoid another peak in infections during the winter that followed.

For many people, it may not have been until the COVID-19 pandemic that we focused on the interpersonal distance we kept between ourselves and other people. We may have had natural preferences for not being too close, but how often did we consciously stop to monitor or assess how close was too close?

The most common time I would stop to think about personal space was when traveling. Before the pandemic, I was lucky to travel to different parts of the world: Africa, Asia, Europe, North America, and Oceania. I did a three-week stint in the -22°F Siberian winter at one point—brisk!

In my travels, one thing that has always struck me is how different countries appear to have very different norms regarding personal space. A simple example is countries where people cram on public transport so tightly that you feel like you might burst out of the door at any moment. It's a far cry from the messaging I've become used to in the UK in recent years.

How we feel about personal space has quite a bit to do with where we come from. In 2017, a large team of researchers conducted a study of nearly 9,000 people to examine expectations of interpersonal distance.[6] The team studied people from 42 different countries around the world—an impressive feat by anybody's standards.

The researchers gave each participant a graphic of the silhouettes of two people standing opposite one another. Below them was a ruler showing the distance between the two people in centimeters.

Each participant was told to imagine that they were one of the people and to use the ruler to indicate how close to them the other person should stand if they were having a conversation. The options were from zero centimeters to just over two meters.

Notably, the participants were asked to make this judgement when considering three different types of people—a stranger, an acquaintance, and a close friend.

The results showed that preferred interpersonal distances during a conversation varied depending on where in the world people came from. Those from Argentina and Peru reported some of the shortest distances for conversations with strangers: an average distance of 75 to 80 centimeters—supposedly similar to a man's average step length when walking.

In contrast, people from Romania and Hungary had some of the largest distances: an average of 130 to 140 centimeters—that's almost like having a broomstick lying flat between you and the other person.

One intriguing finding was that the country's temperature was a critical factor in determining the distance at which people felt comfortable interacting with others. On average, people from warmer countries preferred to be closer to strangers, while people from colder countries preferred to stay further away.

This result has been seen in other research dating back to the 1980s: Warmer climates tend to induce greater social proximity. It also connects to a broader set of research findings that have led some scientists to suggest that hotter weather can create a friendly atmosphere. For instance, further study has found that warmer climates can increase trust between people and the frequency of interpersonal encounters.[7]

The message that people from warmer countries prefer to maintain closer distances from strangers was only half the story of the research.

When it came to perceptions of how close people felt was appropriate for a conversation with a friend rather than a stranger, people from warmer climates were more likely to indicate a preference for being further away from their friends than people from colder climates. In other words, the opposite data pattern emerged in conversations with friends compared to speaking with strangers. In colder parts of the world, people wanted to be closer to their friends when chatting.

One example where a smaller intimate distance was deemed more appropriate for friends was in Romania: People from that country showed one of the largest distances for interactions involving strangers but one of the smaller distances for interactions with someone they were close to. Strangers were encouraged to stay away, but friends could come close.

Why might this be? It has been argued that some of the negative impacts of living in colder climes can be alleviated by engaging in more intimate closeness to individuals we care about. Put a little differently, people in colder countries might get closer to people they care about because it helps them keep warm.

HUGS AROUND THE WORLD

The results above show that people worldwide are very different in their preferences for personal space. I began to wonder how this impacts our tendencies to touch one another. We might imagine that people who prefer to have more distance from others, be it a stranger or a friend, may also like to engage in touch with them less often.

It also might not be that straightforward. According to certain studies, higher temperatures can predict illness occurrence. Living in disease-prone areas of the globe can be related to various behaviors that may reduce the chance of getting sick. It's easy to imagine that someone worried about getting ill might be less likely to want to touch other people. Could this contribute to disparities in views and experiences of close-contact interaction, such as touch, around the world?

I wasn't the only one who wanted to know more. In 2021, a new group of researchers from 64 universities combined forces to investigate whether there are differences in how often people engage in affective touch around the world and what contributes to this.[8] They asked almost 14,000 people from 45 countries whether they embraced, stroked, kissed, or hugged their partner, friends, and youngest child.

The results were telling. Affectionate touching was highly prevalent across all the countries studied, regardless of culture or geographical location: 96 percent of people reported engaging in affectionate touch towards a romantic partner, and 95 percent reported affective touch towards their child.

Certain types of touch were also more common in some forms of relationships than others. Kissing and stroking were more common in romantic and parent–child relationships than between friends. This was a pattern of results found throughout the world and is consistent with the research from Juulia Suvilehto's team demonstrating that stronger ties with other people allow for more touch, irrespective of background.

Alongside these similarities, there were also some differences depending on where people came from. You'll recall that participants were asked about the types of touch they engaged in—embracing, stroking, etc. The researchers used this to identify the diversity of different kinds of touch people experience in their lives. For example, a person who only gave a spouse or partner a kiss each day might have lower affective touch diversity than someone who engaged in hugging, stroking, kissing, and tickling.

Being from a warmer climate was associated with a greater diversity of touch towards partners and friends. So although people from colder climates might prefer to stand closer to their friends, that doesn't necessarily mean that they show as much variety in their tactile behaviors toward each other.

In addition, people from countries that were identified as being more conservative were less likely to engage in diverse types of affectionate touch towards male friends or a partner. An individual's own level of conservatism also decreased the range of affective touch behaviors they engaged in across all relationship types.

You might wonder why this would be. Some suggest that the reasons may connect to early experiences of touch. There is evidence that children from more conservative backgrounds can experience less affectionate touch as they grow up. We know that conservative values and perspectives can transfer from one generation to another. This can include cleanliness preferences, which may connect to views about engaging in behaviors that could pass on germs. It could be that people with more conservative values show less diversity in their affective touch towards others based on factors like these, although further study is needed to confirm this.[9]

Getting older also decreased the tendency to engage in multiple forms of affectionate touch behavior towards a partner, friend, or child. This is consistent with evidence that age can alter perceptions of personal space: Young individuals prefer closer physical closeness, so it makes sense that they may touch more.

It is also possible that a few methodological reasons contribute to why age may have mattered in this study. For instance, people were asked about the time touching their child. You might imagine that young individuals are more likely to have opportunities to touch younger-aged children. In other words, we may touch our own kids more when they are infants than when they are teenagers or adults.[10]

Collectively, these findings build a case for environmental, values-based, and demographic factors that might impact the diversity of affective touch behaviors shown by people around the world towards partners, friends, and children, a sentiment that is echoed in other research studies.

For instance, in 2018, researchers from Arizona State and Pennsylvania State universities aimed to investigate differences in how acceptable Americans with Mexican or European ancestry view affective touch in public. Rather than focusing on romantic partners, this research asked about touching acquaintances in public.

The headline was that, on average, Mexican Americans reported greater acceptability of affective touch with acquaintances in public than European Americans. They were more open to affective tactile exchanges in public from acquaintances.[11]

But this was only part of the story. Views on affective touch in public were impacted by Mexican Americans' self-reported level of acculturation into American culture. In this context, acculturation refers to how much people reported having adopted traits from American culture. Mexican Americans who reported greater acculturation into US culture reported less comfort with affective touch in public settings. The more people felt connected to traits of American culture, the less comfortable they felt about public touch.

This shows us that it is not appropriate to assume people's touch preferences based simply on one grouping factor, like where they come from or what their ancestry is. Just think about your next-door neighbor—sure, you live on the same street, but does that mean you'll behave the same way when touching someone? We must remember that the interaction between touch and culture connects to a range of factors like where a person is from, where they live now, their personality, context, cultural values, age, and relationship status. It's never quite as simple as the cultural stereotypes might suggest.

> **A look back on what contributes to differences in our preference to touch**
>
> There's no doubt that people use affective and nonaffective touch in different relationships worldwide. There are quite a few commonalities between cultures, such as the impact of touch on well-being and how emotional bonds may affect our experience of whether touch is appropriate. There are also subtle differences that can contribute to situations where thoughts and touch experiences may diverge. We must remember that just because two people come from the same background does not mean they will respond to touch in the same way. As we have seen throughout this section of the book, many differences contribute to our preferences for touch, including attachment style, personality, and neurodiversity.
>
> In *Digital Body Language*, Erica Dhawan discusses how being mindful of cultural backgrounds and other individual differences can be crucial to creating global connections.[12] A twenty-first-century collaboration expert, Dhawan explains the importance of showing empathy for different perspectives

in order to create social connections in digital environments. The truth is that many of her suggestions also have implications for physical interactions. If we want to build connections with others, we can start by trying to better understand how differences in our backgrounds might contribute to nonverbal communication through touch. Here are a few insights that I picked up from Dhawan's work that might be important for everyday tactile communication.

1. Discuss differences. If we ignore our differences, they can grow. Provide spaces for talking about differences in our thoughts and feelings about touch.
2. Speak up about your preferences. If we want to increase our understanding of how we differ in our thoughts and feelings about touch, we need to feel comfortable expressing these. It is probable that by modelling the behavior of speaking up, we will encourage others to do the same.
3. Be adaptable. Even if you are interacting with someone you've known for a while, remember that feelings towards touch can change. Be respectful of each person's boundaries.
4. Avoid mapping your preferences onto others. As we have seen, several individual differences can change how we respond to the same tactile experience. Psychological science has taught us that we often overestimate the number of people we think are like us and share our preferences; this is called the false consensus effect. We must be careful of this bias in thinking and consider how other people might feel when choosing to touch them.

PART 4

TOUCH MATTERS

CHAPTER 8

Social Touch: The Hidden Secret to Effective Teamwork

It was October 2008, the opening game of the Los Angeles Lakers' season. The Lakers won. They would become NBA champions that year, their first championship for seven years before back-to-back titles.

Most people, like me, watched the season-opening game and those that followed through the eyes of a fan. However, a group of researchers at the University of California, Berkeley, watched these games differently. They were coding touch between teammates, including the number of fist bumps, high-fives, chest bumps, and hugs, as part of an intriguing experiment that sought to see if touch between teammates could predict future success.

By painstakingly counting the frequency of early-season touching from a single game for every NBA team, the researchers were able to see how teammate touch was related to performances later in the season. They found that teams with higher levels of touch early in the season showed increased cooperation and better performances later in the season. Early-season touch predicted better performance for an individual player and their team.[1] The researchers discovered that touch appeared to be linked to greater teamwork and cohesion, which in turn seemed to aid performances.

*

Sport is one of the most common places to see touch between individuals openly demonstrated in society. Whether celebrating success, encouraging, or commiserating, I'm sure we can all think of examples where we have seen athletes, coaches, and fans using touch as a form of nonverbal communication.

For the Liverpool Football Club fans amongst you—or even if you just like Premier League football—just think about the Klopp hugs.[2]

In case you do not know, Jurgen Klopp is a charismatic manager. You will regularly see him animated on the touchline at full throttle. He is famous for many things in football, including winning Liverpool Football Club's first league title in three decades, manager of the year awards, and multiple cup wins. But there's another thing that he is known for—his hugs.

Neck–waist hugs, crisscross hugs, bear hugs, from-the-back hugs: The internet is full of montages of Klopp hugs. Klopp himself places great weight on these in the context of the sense of unity and camaraderie in his squad. After winning one match, he famously sought out every player for a bear hug. Some reporters asked him why.

"I enjoy it more," he replied. "I'm really demanding to be honest, and I really want a lot of them. When you can really see how they fight, with the last drop of fuel in their machine . . . that's the most easy thing to do [hug them]. This is what makes it more enjoyable for the players. Having something like this in the dressing room and seeing the players there all smiling but tired . . . that's really nice."[3]

As it turns out, Klopp's hugging approach is a smart one: Shared tactile nonverbal communication has been linked with team cohesion and positive psychological momentum across a range of sports, including American football, soccer, netball, hockey, lacrosse, swimming, and basketball.

Some athletes also report that touch between teammates during competition can help them get in the opposition's head. In this way, touch doesn't just give a competitive edge by building connection; it is also perceived to demoralize the opponent. I should add here that we don't know if touch does demoralize the other team. But it's easy to picture that if you see another team looking more cohesive, you might worry that they are harder to beat.

Not surprisingly, professional sporting teams now take an interest in touch between their players and coaches. Eight years after the Berkeley study described at the start of this chapter, at least one NBA team, the Phoenix Suns, developed their own "high-five stat" to track how often player-to-player hand slaps occurred during a game.[4] The aim of this was to help monitor and build team trust and communication. That year the Suns finished bottom of the Western Conference. Five seasons later, they were crowned Conference champions.

Of course, I'm not trying to suggest that the high-five stat was the reason for the Suns' resurgence. Success has many parents after all. Yet when coupled with data from other sports, there is quite a compelling case for the potential of using positive and welcomed touch to fuel sporting success.

As someone who plays and follows a lot of sports, I wanted to know more about touch and sporting success. In my search to better characterize these relationships, I stumbled across a fascinating study led by researchers from California State and McGill universities, which examined how different types of positive tactile communication were used as a coaching strategy in a successful women's basketball team over two years.

Although the researchers did not name the team, they were known to have won multiple conference championships and participated in national championships for three consecutive years before the study started. The team won two more consecutive conference championships during the two-year study. In short, they were pretty darn good: the kind of team where when a trick or two is revealed about their success, we should probably listen up.

Over the years of working with the team, the researchers recorded videos of the practice sessions involving coaches and players. They coded 18 different types of positive tactile interaction that took place. These included behaviors like high-fives, fist bumps, and placing hands on top of each other. They also examined various kinds of positive tactile interaction during competitive matches, creating hours of extensive coding of touch in a sporting environment. The researchers then interviewed team members, asking their thoughts about these positive tactile behaviors.

When asked about the outcome of positive touch from their coaches, all the players reported that it had a beneficial impact on their individual and team performance.[5]

A positive touch from the coaches also resulted in the athletes reporting improved emotional states, cultivating feelings of trust. Touch strengthened interpersonal relationships, particularly for new team members, for whom positive tactile interaction was thought to help promote team unity and a sense of belonging.

One of the intriguing aspects of the research was that most of the positive touches shared between the team were brief. The most common forms of touch were exchanges lasting for one second or less, such as a quick pat on the shoulder. These short bursts of touch occurred roughly 80 percent of the time or more in practice or competition. They may not have always involved someone that a team member had a close bond with. They did, however, still carry meaning. Brief social touch exerted a powerful impact on team cohesion and performance.

TIPPING POINT

Touch in sports is just one example of how a brief social touch can impact people's behavior and feelings. Many other examples occur throughout society right in front of our eyes.

Some of the clearest demonstrations of the impact of brief social touch on behavior come from social and consumer psychology research. Over decades, scientists have studied how subtle social touches impact evaluations of people and surrounding environments. Extensive research has shown that even the simplest of touches can change how we spend our hard-earned cash.

In 1984, researchers from the University of Mississippi and Rhodes College took a trip to a restaurant. They weren't there for your standard trip to the family diner or a romantic dinner for two. They went to study how different interactions between waitstaff and restaurant patrons contributed to differences in tipping—where an extra sum of money is given by a customer to a staff member for the service they have performed, above and beyond the price of the service itself.

In the study, waitstaff briefly touched some customers when they returned their change after the bill was settled. The touch was either on the customer's hand or shoulder.* We probably wouldn't give a second thought to this type of innocuous gesture if it happened in a restaurant we were visiting.[6]

The outcome of this short touch, however, was powerful. Tipping rates were greater when the waitstaff touched the customers than when they did not. People were willing to give

---

* Hand touch saw the waitstaff touching the customer's hand twice for 1.5 seconds. Shoulder touch saw them place a hand on the customer's shoulder and hold it there for 1 to 1.5 seconds. The waitstaff did not smile when they spoke and simply used the words "Here's your change" in a consistent neutral tone.

more additional money for the service they received after being touched.*

In the no-touch condition, they gave 12.2 percent of the bill. In the shoulder touch, they gave 14.4 percent, and in the hand touch they gave 16.7 percent of the total bill. If my math is correct, and we assume that a staff member might wait on ten tables a night for five nights a week, with a spend of $50 a table, this would be close to $6,000 of extra income a year.† That's a handsome sum simply by adding a short touch along with the bill.

Following this seminal work on restaurant tipping in America, other investigations throughout the late 1980s and 1990s have pointed to a similar pattern of data.

Some extended the results to different hospitality settings and customer behaviors. For instance, researchers at Virginia Commonwealth University found that men and women consumed more alcohol when touched by their serving staff compared to when they were not.[7] Their behavior was influenced simply by whether they received a touch on their shoulder for a moment while being asked if they wanted a drink.

Other researchers looked to see if the results of greater tipping following touch would extend outside of America. Researchers from the Université Bretagne Sud found that like their American counterparts, French customers tipped their server more when touched during a dining experience.[8]

What caught my attention in the French study was that it was conducted at a time when tipping could be considered unusual, because legislation mandated that a 12 percent service charge be included on the menu. Service costs were already included

---

* Other factors that could influence the results were also controlled for in the analysis. These included: weather, alcohol consumption, day of the week, and number of people in the dining party.
† Assuming 48 working weeks.

in the price. This means that even after adding a tip, people still gave more when touched. All this makes clear that relationships between socially acceptable brief touches and positive evaluations of people and surroundings exist in different parts of the world and in different economic situations.

I was shocked by the effects of touch on tipping. I began to wonder if brief touches can impact other retail settings. It turns out that a similar pattern of results has also been found in shops, where appropriate touch given by front-line employees has been found to extend customers' shopping time.

In the 1990s, Jacob Hornik of Tel Aviv University reported multiple experiments examining how touch impacts consumer responses. In one study, shoppers were approached as they entered a large bookstore. Each shopper was given a catalog containing information about discounts and items in the shop. Notably, some customers were touched lightly on the upper arm as they were handed the catalog, while others were not.

Those who were touched spent almost 63 percent more time in the store than those who were not touched. They also rated the store more favorably and spent nearly 23 percent more on products.[9] I know that a lot of shopping takes place online these days, but those are some impactful numbers.

Incidentally, there are now studies looking at how viewing touch can impact purchasing intentions as well. A brain imaging study led by researchers from Shenzhen University found that watching other people briefly touch products increased viewers' intent to buy the items.[10] In a modern world full of online shopping and media ads, we clearly need to be mindful of the impact that both physical touch to ourselves and observed touch to others can have on our decisions to buy things.

All this makes me wonder: If brief touch can lead to bigger tips or more lavish customer spending, does anyone try to take advantage of this commercially? Perhaps it is just something about me, but I can't recall the last time a waitstaff member touched me on the shoulder when I ordered a drink. Surely it would be in their interest to do so. The data suggests that I would drink more and tip them more generously. Perhaps they do touch me briefly, and I don't notice. Or maybe most customer service staff, and their managers, don't know about these findings. If touch can lead to positive effects for the person initiating it, what would happen if staff were actively instructed to touch?

It turns out that how service staff feel about engaging in touch in customer service settings is important to the outcomes of using touch in such settings. In 2020, scientists from the University of Iowa examined what happens if restaurant waitstaff are told they must touch a customer briefly during their interactions. In other words, if they are given a directive to touch to help raise revenues.

The outcome was not straightforward.

When staff were instructed to touch customers, they expected the customer to feel worse about the touch than they actually did.[11] Staff members were concerned about appearing too personal when touching the customer. Even if the customer was fine with the interaction and left a higher tip, the staff did not perceive this to be likely to happen. The staff felt more uncomfortable than they needed to be. What made matters worse was that when staff members felt this way, they became less likely to engage in future interactions with customers. Trying to enforce touch backfired. Perhaps touch between individuals must be spontaneous and organic to be effective—another nuance to consider.

THE PERSUASION ROBOT

When I was a teenager, I had a weekend job working in a local clothing store. My Saturdays were filled with scanning goods and packing shopping bags. The endless beep of the checkout. The constant buzzing of a bell would ring to let us know that the line had grown too big, so more people were needed at the cash registers.

One day I was working at the checkout desk. I'd scanned and bagged several items for the customer, and it was time for them to pay. At this point, the customer—a young mother with a baby—started fumbling in her purse. She was getting anxious and carefully counting her money again and again. With a brief touch on my arm, she told me, "I am so sorry, I'm twenty pence short. Will that be okay?"

I paused for a moment. What should I do? It was only twenty pence, but it wasn't my twenty pence to decide upon. There was a big line and no manager to ask. I had to make a choice. Say no and encourage the woman and her child to come back. Say yes and face the consequences of my cash register not balancing at the end of the day.

Seconds passed, but it felt a lot longer to a socially anxious teenager. I looked at the woman. I looked at her baby. I needed to make a decision. I decided to let the customer pay me twenty pence less than she needed to.

My other colleagues were shocked. "You can't do that!" exclaimed the girl at the checkout next to me. I felt embarrassed. But I'd made the decision, and it was done. I moved on to the next customer, hoping I wouldn't have to do any more favors that day.

Although I was okay with my decision, I will admit that I later ruminated over what had happened. I wondered why I'd agreed to let this twenty-pence difference—which clearly upset my colleagues—go.

It was some years later that I came across an experiment that helped me understand aspects of my decision-making process that day.

This experiment involved bus drivers in a French town with a population of roughly 40,000 people. These drivers were used to driving around the town. Making stops en route. Greeting customers. Taking money from them when they bought their tickets. The day the researchers conducted their experiment was no different. Apart from one thing: On that day, research assistants boarded the buses.

The researcher asked the bus driver for a ticket. They spoke briefly with the driver—a regular customer interaction. Then it came time to pay. At this point, the research assistant began to look perplexed. They searched through their wallet. Rummaged in their pockets. Looked in their wallet again. They were just short of having the right amount of money to pay for the ticket.

They explained their situation to the bus driver. Could they still ride the bus?

Importantly, the driver was unaware that they were part of an experiment—to them, the research assistant was just like any other customer on their route. Even more importantly, the request involved a brief touch for some bus drivers. For others, it did not.

When touched by a female research assistant, 60 percent of bus drivers agreed to the request to let them ride for a reduced fare. Only 35 percent agreed when no touch was involved. When touched by a male research assistant, 25 percent of bus drivers accepted the request. This compared to 10 percent of drivers agreeing to a male research assistant request when no touch was involved.[12]

The bus drivers accepted the customer request more favorably when touch was involved, albeit with nuances connected to the gender of the person doing the touching. A simple behavior, a powerful impact on willingness to help another person.

It's not just in customer interactions where brief forms of social touch can increase friendly and prosocial behaviors. Simply including brief social contact with a request—such as a touch on the arm—can increase our willingness to take part in surveys, to give money to charity, and even to look after a stranger's dog for ten minutes while they pop into a shop.

Different scientists across different decades have found this pattern of data. One classic study in the 1970s found that people were more likely to give back a coin left in a phone booth if the previous caller touched them when they left the booth than if they did not. Nearly three decades later, a 2007 study in France found that people were more likely to give away a cigarette if the request was accompanied by touch.[13]

These findings show us how rudimentary touches that we encounter in everyday social interactions can provide a starting point for cooperation and compliance. A brief persuasive touch can exert a strong impact, even between people who are otherwise strangers.

When I found out about these results, it had me second-guessing all those times that I'd agreed to requests from people over the years. The customer in my old job; the charity worker who persuaded me to donate on the street; the stranger who convinced me to buy them a beer. Did they all know about the magical powers of touch? Could touch have been purposely used to encourage a positive response? Is that even ethical?

One domain where we might be willing to see the power of persuasive touch used is in supporting healthy behaviors. In 1980, researchers from the Department of Health and Nutrition Sciences at Brooklyn College examined the effect of gentle touch on eating behavior in elderly patients who struggled to provide for themselves. Usually, to encourage the patients to feed themselves, staff would talk to them, verbally encouraging them to eat more. In the experiment, some staff added a gentle touch to the patient during this exchange. This small addition of touch contributed to greater nutritional intake. The patients ate more.

Even more impressive was that the positive effect of touch on eating behavior was found to last for up to five days after the first tactile contact. Touch acted as a helpful nonverbal cue to promote a long-lasting positive health behavior.[14]

The relationship between brief social touch and compliance has been found in other settings. In some research conducted in classroom settings, a short neutral touch on the forearm from a teacher to a student has been shown to improve the likelihood of students volunteering. It also impacts perseverance.[15] Researchers from Oakland University in Michigan found that students are not only more likely to take part in a difficult task after being touched, they stick with it for longer too.

Even touch from a robot has been shown to impact how likely we are to engage with requests. In 2021, researchers in Germany published a study investigating how university students behaved during encounters with a humanoid robot that either did or did not touch their hand during a counselling conversation.[16] The study, led by Laura Hoffmann and Nicole Krämer, involved students interacting with a robot that looked a bit like a miniature version of Baymax from the Disney movie *Big Hero 6*.

The students were allocated to one of two groups: One group just talked with the robot, while in the other, the robot reached out and touched them, patting the student's hand three times before continuing the conversation. Think about it for a moment—do you think you'd be more likely to comply with a request from a robot if it touched you or just spoke with you?

This is precisely the question that Hoffmann and Krämer asked. The request from the robot was for the students to join a course. The results showed that 81 percent of students answered "yes" when the robot touched them. Only 59 percent answered "yes" when the robot did not touch them. Touch from this little robot helped persuade people to comply.

In a society where interactions between humans and robots are likely to become more commonplace, these findings have powerful implications. Some hospitals and nursing homes already use robots to support interactions with patients. Could persuasion robots help encourage patients to engage in positive health behaviors like eating?

What if a customer service robot touched you on the way into a shop—do you think you'd comply more with a request to try a new product? Should this be allowed? We don't know the answer to these questions. Still, as we'll discuss more in our final chapter, we can expect more insights into human–robot tactile interactions in the future.

THE COVID TEST

Before moving on, it's worth pausing to consider another implication of the findings we just read. All the studies above point to people feeling more comfortable with a request following touch. In some cases, like encouraging people to eat, that can be a good thing. But can feeling more comfortable after social touch ever have an unintended negative consequence?

Insight into this question was recently provided by researchers from Southwest University and the University of Economics and Law in China.[17] They conducted a study during the early stages of the COVID-19 pandemic that examined how brief social touch impacted engagement with preventative measures to support public health.

To understand this study, I'd like you to picture a scene where you arrive to participate in an experiment about time management. Perhaps you'd arrive early. Maybe you'd arrive late. Irrespective, you arrive and sit down in a waiting room.

The researcher comes to meet you. You greet one another—safely distanced—and the researcher explains that they will take you to another room to complete the experiment. They tell you that there will be hand sanitizer on the table as you enter the testing room. You should use this to safeguard your hands against germs due to the impact of COVID-19.

You enter the room and see the sanitizer. My question for you is: Do you think you would remember to sanitize your hands?

In the study, roughly 75 percent of people did sanitize their hands when entering the room. They remembered and complied with the request.

Now for the twist. The same interaction happened for another group of people. Meet—greet—get told to sanitize—head to the testing room. Yet there was one crucial difference. During the explanation about the sanitizer, this group of participants were given a comforting pat on the shoulder by the greeter. This touch lasted for no more than two seconds. The participants then went into the testing room to complete the survey, the critical outcome measure being: Would they sanitize their hands?

Based on the findings we've discussed so far, you might expect that they would. After all, brief touch increases compliance, right? Wrong. In this study, only 56 percent of people in the touch interaction group sanitized their hands. This was even though their memory of the instruction was just as good as those who were not touched.

The group of participants who received the comforting pat on the shoulder were almost 20 percent less likely to sanitize their hands upon entering the room. Why would this be?

One reason put forward by the researchers was that perhaps the touch from the experimenter hinted that it was okay to violate preventative measures. After all, they had already engaged in touch, which was somewhat taboo in pandemic-related behaviors.

To examine this theory, they conducted a follow-up experiment with different participants a month after the first study. They took the same approach but adjusted the type of touch. This time, the experimenter either patted the participants on the back, shook their hands, or did not touch them. Suppose a pat on the back hinted that breaking pandemic rules was okay. In that case, you might expect that you'd see something similar for a handshake—arguably a stronger violation of preventative measures to stop the spread of a virus.

Once again, participants who were patted on the back were less likely to sanitize their hands when entering the room, with only 59 percent of people engaging in the behavior. In contrast, 79 percent of people sanitized their hands following the no-touch interaction. Importantly, 81 percent sanitized their hands following the handshake. It appeared that it was not engaging in touch in general that was driving the behavior change. Instead, something special happened in relation to the pat on the shoulder that people received.

What might this be? An intriguing finding from the research was that the main factor that helped explain why the participants repeatedly broke away from preventative measures following the brief shoulder touch was their perceived sense of security.

People in the shoulder-touch group reported higher levels of security than those exposed to other types of interaction before entering the room. The more secure the individual felt, the more they breached the rules. A brief comforting touch from the researcher made people feel safer, which contributed to a greater likelihood of breaking hygiene rules, despite having been explicitly asked not to do so.

As with so many examples of tactile interaction, this study offers a fascinating insight into how complex all forms of social touch can be. Despite decades of research suggesting that brief touch increases compliance with requests (even when that touch comes from a robot!), the unique circumstances of a global pandemic turned that finding upside down. This study offers intriguing information on how situations, circumstances, and environment might impact how social touch influences human behavior.

Whether it is another person or a robot, in education settings, shops, or nursing homes, it is evident that a simple touch, like placing a hand on the shoulder for a fleeting moment, can influence how we perceive and act with others. Hidden in plain sight, these subtle but important influences of touch impact our daily lives in ways we might never stop to think about.

THE BEAUTIFUL PEOPLE

One of the things that the study on social touch and hand sanitizing shows us is that, like all forms of touch, the outcomes of brief social touch can be more complex than they first appear. It's not just COVID that demonstrates this. Even before the pandemic, many other elements had been shown to impact the outcomes of brief social touch. One aspect of paramount importance relates to whom the person touching us might be.

Several of the examples showing positive outcomes of brief tactile interactions that we discussed earlier involved situations where a short touch might be considered an acceptable part of the exchange taking place. Touch from our restaurant server or from a member of staff as we enter a store might fit our expectations or social norms about what can occur in customer service settings. Therefore we might find it acceptable for touch to happen in these interactions.

It's easy to imagine that if we don't find the touch we receive acceptable, it might lead to a very different outcome than more tips for the waitstaff member. If this touch isn't welcomed, it's unlikely we'll stick around for long or reward the behavior. Science backs up this commonsense assumption.

In 2011, Brett Martin of Queensland University of Technology showed that accidental touch from another customer resulted in people spending less time in stores than when no touch occurred.[18]

The study saw people being accidentally touched while they were browsing products. To illustrate, picture someone bumping into you by mistake when you are trying to pick the best fruit in a supermarket.

Martin compared this type of interaction with a situation where the other customer brushed past the participant while they viewed a target product. In other words, they got close, but they did not touch.

An accidental touch from a stranger had several adverse outcomes on the customer's brand evaluation and purchasing intentions. These included the customer being less likely to buy the product and reducing their shopping time. Even before the pandemic, it seems that many people were not fond of close encounters with strangers when doing their shopping.

It does not take much digging to find a few other factors that can influence how people respond to brief social touch from people they do not know. One major element affecting how social touch influences our behavior is our thoughts about in-group and out-group members.

Social scientists define our in-group as a social group that we identify with belonging to. In contrast, an out-group is a group that we do not identify as being a part of. We can identify with several different in-groups: family, communities, ethnicities, genders, nationalities, religions, political affiliations, or even sports teams. For instance, I'm a Liverpool Football Club–supporting British-Iranian psychologist. In some settings, I may identify with just one of these groups, but in others, multiple.[19]

It's been shown that having this extra information about the groups I identify with might change how you respond to a brief tactile exchange between us if we were to meet.

For instance, although we know that brief touch can lead to better impressions of people, brief touch from an out-group member can take longer to exert a positive impact than a touch from an in-group member.[20] Essentially this means that if you happen to be a fan of a different football team than me, you'd likely respond differently to touch from me than from a fan who supports your team. If we were to touch each other at a match on our first meeting, you'd be less likely to have a positive impression of me compared to if I supported the same club as you.

Individuals can also show differences in interpersonal touch response based on their perception of other people's traits and physical appearances. Some people indicate more discomfort with interpersonal touch if they believe that the person touching them has a health difficulty or negative criminal history.

One reason for this is possibly a fear—often misguided—of some kind of contagion. It connects to a broader phenomenon that sees us responding differently to social touch based on how attractive we perceive the person touching us to be.

Put simply, the more attractive we find a stranger, the more likely we are to experience their touch as pleasant.

One demonstration of this can be seen in how we respond to gentle stroking. Studies demonstrate that people show more pleasurable feelings and changes in heart rate if they are shown "attractive" faces rather than "unattractive" ones when their arms and hands are stroked.

In one study, faces were shown on-screen while people were stroked by someone behind a curtain. The participants were asked to imagine that the person they saw on the screen was the one behind the curtain touching them. Each face was taken from a database of faces previously rated for attractiveness by a different group of research participants.* Perceived attractiveness impacted

---

* The participants in the experiment also rated the perceived attractiveness of the faces after the experiment was finished to ensure that their judgement

the feelings of pleasant touch: The more attractive someone was, the nicer the stroking, even if the sensory qualities were similar.[21]

The finding that attractiveness can change how we perceive touch is consistent with wider research in social psychology that suggests people tend to have a "beauty-is-good" stereotype.[22] That is to say, we don't just tend to prefer touch from attractive people; we can also have a bias to assume that attractive people are smarter, more trustworthy, and more competent.

One argument for why we prefer touch from attractive people is that we tend to think of attractive people in a more positive light—we are biased to favor beautiful people.

Remarkably, some research even suggests that caregivers can be less affectionate to their children if they perceive them to be unattractive.[23] I don't know of studies that have connected this specifically to touch. But this is quite startling, given the importance of touch and affection in our lives. It speaks to the importance of finding ways to overcome biases against less beautiful people. One way that scientists suggest we can do this is by raising awareness that our biases exist in the first place—just reading this paragraph is a start.

We should also be mindful that attractiveness is not a static concept. It can change with exposure and as we learn more about other people. Some research suggests that being touched by a stranger can also increase our perceptions of how attractive we think they are.[24] So, although we might respond better to touch from a beautiful stranger, our perceptions of their attractiveness may also be influenced simply by being touched by them.

The reasons for this aren't clear. But it is worth noting that considerable research from social psychology has indicated that as we have more exposure to others, there is a greater likelihood of attraction to them.

---

of "attractive" versus "unattractive" was consistent with the previous ratings.

Relatedly, some research shows that simply knowing someone likes us can impact how attractive we find them.[25] Could it be possible that touch from a stranger makes them more appealing to us because we interpret the behavior as a signal that they like us? Does the type of touch matter?

There is limited research on this topic, so our understanding of why this might be the case requires further investigation. The primary research on which it is based comes from a single study in the 1970s. It would be interesting to see whether the results still hold today, and why. For now, perhaps hold off on touching strangers as a strategy to improve your own attractiveness rating until the science has more answers.

> **Reflections on brief social touch**
>
> All these studies illustrate a similar point: The outcomes of brief touches on behavior can be powerful. But they can also be influenced by context and individual differences. Think back to all the differences we've discussed so far that can impact upon touch: age, attachment style, culture, gender, neurodiversity, and personality. These are just a snapshot of factors that might affect how a brief social touch can be interpreted differently—a difference that contributes to why it's not always the case that interpersonal touch will lead to people spending more money or giving positive evaluations. The outcomes of even the briefest social touches are influenced by our expectations, desires, and perceptions of touch in each moment. Here are some key messages from this chapter about the impact of brief touches in different social settings.
>
> 1. Brief touches can impact spending. Whether giving tips, spending more time in a store, or making purchasing decisions, a welcomed brief touch has been shown to

impact consumer behavior across several domains (hospitality, retail, and even online shopping).
2. Brief touches can impact helping. Whether sharing our time, giving to charity, or lending a hand to someone who needs extra change for a ticket, a brief welcomed touch has been shown to impact helping across different continents.
3. Brief touches can impact compliance. Whether in care settings, educational environments, or robot–human interactions, a brief welcomed touch can be persuasive and lead to more compliance with requests.
4. Brief touches can backfire. Despite the power of appropriate brief touches, they can have unintended consequences. For instance, if staff are instructed to touch customers but do not feel comfortable with this, they may interact less often with the customers.

CHAPTER 9

Do Touch, Don't Touch: The Murky World of Touch at Work

Picture the scene: You're working in an open-plan office, surrounded by your colleagues. Some have headsets on, talking away online. Others are absorbed in documents that are open on their screens. Working in this big back-office team, you are writing up notes for an important upcoming meeting, trying to ensure the necessary paperwork is in place.

Amidst all this industry, your boss walks in. This boss is a great person, well respected and highly capable. They are someone you've known for a while. Your company tends to promote from within, and you joined the company around the same time.

They greet the team and get updates on how work is progressing. There is a quick pat on the back for one colleague and a thumbs-up to another. The boss finds out that you've made great strides in preparation for the meeting. Impressed, they rub you on the shoulder approvingly, say well done, and then move on.

What I've just described might seem like an average day at the office for some people. Even if this situation doesn't happen where you work, you may not have batted an eyelid at my description of the situation. But for others, a few things may have stood out. A pat on the back or a rub on the shoulder might be commonplace in some parts of life. Still, to many people, they sound inappropriate and inconsistent with expectations about how workplace interactions should occur.

Touch at work is a topic that is often divisive and a cause of disagreement. The mere mention of workplace touch can lead many to immediately think of extreme examples of inappropriate touching, such as disturbing interactions that define international movements like #MeToo and other large-scale misconduct involving sexual harassment in professional settings through unwelcome touch. These undeniably wrong behaviors have rightly emphasized how physical boundaries have too often been intruded upon in professional contexts.

But beyond clear-cut examples of sexual harassment, other situations can make us ask how far is too far when engaging in touch in professional settings.

When I worked in retail, I had a very friendly, tactile manager. They would often touch me on the shoulder or squeeze my arm when working. We never established whether I was comfortable being touched in this way; it just seemed that my manager could not interact with people without doing so. At the time I didn't think too much of it, but in hindsight, I can see how these behaviors could have been unsettling.

Part of the reason for this is that any form of touch can become inappropriate and intrusive without shared consent. Unwanted touch at work can risk the receiver feeling uncomfortable, undermined, and less able to get on with their tasks.

Some research suggests that people find touch more harassing than verbal behavior.[1] Touch on a coworker's face is often viewed as one of the most inappropriate workplace behaviors. And I think we can all agree that even though we wouldn't necessarily define touching someone's face as sexual harassment, it implies a kind of intimacy that is not welcome at work—especially when uninvited.

These and other potential downsides of touch in professional settings have rightly brought about what is known as "touch skepticism." A survey of almost 2,000 people by a major UK recruitment agency in 2019 suggested that 76 percent of people wanted to see physical touch reduced in the office.[2]

My own research team has also found that amongst a sample of roughly 17,000 employed UK adults, nearly 72 percent agree with the statement "I'd feel uncomfortable if a colleague touched me on the shoulder in public." This contrasted with 13 percent who disagreed with the statement and 15 percent who did not mind either way.[3]

It is perhaps not surprising that some organizations have even gone so far as to introduce no-touch policies. For instance, some schools have strict no-touch rules that mean that students can be sent to the principal's office for hugging, holding hands, or even high-fiving each other.

Yet among these reservations towards touch in professional and educational settings, others argue in support of touch at work. They contend that simply removing all forms of physical contact could mean that we lose opportunities for the positive effects of touch between colleagues. Those who support touch at work say that appropriate touch can help reduce anxiety and stress. It can comfort and reassure. When used well, it helps to build and maintain connections between individuals and teams. All these benefits are potentially positive features in some professional contexts. And all are reasons why some researchers have explored whether there are times when the benefits of touch in professional contexts might outweigh the costs.

THE APOLOGIST

Let me level with you. When I first started writing this chapter, I was skeptical that I'd be able to find convincing evidence that the benefits of touch at work could outweigh the costs. In fact, there is a surprising lack of direct research into the area of appropriate workplace touch to provide scientific validation.

I started thinking back through some of the key messages we've come across in this book already. Touch can help teammates in sports: Okay, that's one workplace where touch could be good. But even here, some very well-known examples exist where people have abused their license to touch.

For instance, take the troubling stories of sexual abuse in US gymnastics that have come under the spotlight of public attention since 2016. They involved hundreds of people and reportedly spanned over two decades.[4] I'm not sure that the benefits of touch on team performance could ever outweigh these awful negatives.

I began to think that rather than consider ways in which engaging in touch might be worth the risk, I needed to reframe the question. I needed to consider when touch might be important in professional contexts. The complex question of whether we should or should not allow people to touch at work might be easier to address once we have a better grasp of the beneficial ways that touch can be expressed between coworkers.

So, in what situations might socially appropriate touch benefit people at work? A straightforward example might be to communicate.

To help explain this, I should mention that people are surprisingly good at identifying different types of emotions from others through touch. When people are touched on the arm by a stranger, they can correctly distinguish anger, fear, disgust, love, gratitude, and sympathy from that touch alone. In some cases, people also prefer to use touch to identify emotions compared to visual signals or body movements.[5]

To show you what I mean, in one experiment, over 200 participants from a large public university in California volunteered for a study that saw them touching strangers. The participants sat at a table opposite another person, with an opaque curtain blocking their view of each other, aside from a bare arm (from the elbow to the hand) placed through the curtain. Imagine that, sitting in a room with nothing but a random person's arm stretched out on a table in front of you—all a bit *Evil Dead*!

The participants were shown words describing different emotions. They were asked to think about the emotion—for instance, "anger"—and then use touch to express it on the other person's arm.

Importantly, the person receiving the touch did not know the emotion being conveyed to them. Nor could they see or hear any other signals from the person touching them. All they had to judge the emotion was the touch they felt.

Those receiving the touch were surprisingly good at perceiving emotions expressed by touch alone, in particular, anger, fear, disgust, love, gratitude, and sympathy. (If you are wondering what anger might have felt like, the most common tactile behaviors used to express it were hitting, squeezing, and trembling.)

Anger might not be great in the workplace, but some other emotions expressed by touch are. Gratitude is particularly important. Just pause for a moment and think about the last time someone you worked with stopped and gave you a genuine expression of thanks for your work: How did it make you feel? Chances are, you felt pretty good and appreciated.

Expressions of gratitude are known to enhance relationships and performance in the workplace. Feeling valued at work relates to being more likely to have higher engagement, satisfaction, and motivation.[6]

What's more, expressing gratitude to a teammate can improve performance in stressful situations. In 2022, researchers from the University of California San Diego's Rady School of Management established that simply asking teammates to thank each other before performing a collaborative activity improved their cardiovascular response to the stressful situation.[7] Those teams with improved cardiovascular response showed increased attentiveness and confidence. All these factors contributed to workers performing better at the task they were completing.

In other words, expressing gratitude was positive for people working together. If touch can help us express gratitude, then arguably, when done right, a brief touch to express gratitude might have some value in the workplace. Still, we should also recognize that gratitude can be expressed without touching too. Appropriate touch might add to this in some situations, but that doesn't give free rein to go around touching people at work without their consent: For many, any form of workplace touch is never appropriate.

Another example where an appropriately applied workplace touch has been shown to be important is establishing the emotional sincerity of the people we interact with.

To help explain why this matters, think again about a colleague expressing their gratitude towards you at work. Sure, someone saying thank you is positive, but it's likely that to truly experience feelings of appreciation, we'd need more than simply hearing the words. We need to believe the other person means what they are saying. To convey emotions like gratitude in a way that lands positively, we need to feel that the sentiment behind it is sincere. Put more simply, I have to believe that you mean it when you tell me that you are thankful for something I've done.

Touch, it turns out, plays an integral part in how we perceive the sincerity of some behaviors in the workplace. When people are asked to assess videos of one staff member apologizing to another, they rate the apology as more sincere when touch is used.

This was shown in an experiment published in 2011 in which university students were asked to watch a two-minute video of a supervisor taking credit for another staff member's idea. Afterward, the supervisor apologized to the employee.

Importantly, different students saw slightly different apologies. In one exchange, there was no touch between the supervisor and the employee. They simply spoke and went on their way. In other videos, there was touch. For instance, people saw the supervisor shake hands with the employee at the start and end of the apology. Aside from these differences in touching behavior, the content of the apology was the same.

After watching the various versions of the video, the different groups were asked to consider whether they thought the staff member would forgive the supervisor. They were also asked if they felt the supervisor was supportive and sincere in their apology.

The results showed that the perception that the supervisor had genuinely apologized was higher in students who viewed the handshake apology video than in those who saw the no-touch video.

Some students also viewed an apology involving a handshake plus a pat on the back: Here again, they rated the supervisor as more supportive than students who viewed the no-touch video.[8]

Simply put, touch contributed to perceptions of the sincerity of the apology. The manager appeared to be more supportive and sincere by using appropriate workplace touch when apologizing.

Why is this important in relation to how we may wish to use touch in workplace interactions? Simple—saying sorry for mistakes made is an important part of life at work. During teamwork, people can occasionally say or do things that offend colleagues. People can get things wrong. These situations can, of course, lead to friction, especially if they are not resolved. Apologizing for mistakes we've made or offense that we've caused can help reduce these tensions and resolve conflicts that would otherwise risk damaging working relationships.[9]

When providing a genuine apology, touch can help that apology land more effectively. It may not always absolve the misdemeanor—no reassuring pat on the arm will compensate for someone who routinely steals your ideas. But it might be helpful in some workplace situations.

THE ONE WITH THE ULTIMATE FIGHTING CHAMPION

If you happen to be a fan of the TV show *Friends*, you might recognize the title of this section. It is the penultimate episode from the third season—one of my favorite seasons, by the way.

This episode saw Monica's boyfriend Pete decide that he would become the Ultimate Fighting Champion. It is also famous for a workplace touch interaction that takes place between two other characters, Chandler and Doug.

For those who do not know, Chandler is one of the six main characters in the series. The show has a running theme about how he dislikes his well-paid office job.

In the Ultimate Fighting Champion episode, viewers see another side of his work. Chandler gets a new boss named Doug, and we see the two men interacting in the office. It all seems like a regular office exchange, until the final moments. Doug has a very unusual way of congratulating Chandler on his excellent work: He slaps him on the butt, leaving Chandler bemused and the viewers laughing.

One of the things that stuck with me about this episode was how differently Chandler and his colleagues felt about Doug's butt slapping. Chandler was, in my opinion, rightly horrified. The slap didn't just happen once. It happened time and time again.

His colleagues, however, were jealous. Doug only ever slapped Chandler's backside. Chandler's colleagues felt that, clearly, their boss was overlooking their good work.

I won't go into all the details that follow—you're not here for a full breakdown of *Friends* Season 3, after all. But I wanted to use this episode to pose a set of questions to you. Suppose we accept the argument that sometimes touch can carry benefits for workplace interactions. In that case, how should we tailor touch to suit the people we interact with? What is the best way to approach the murky world of touch in the workplace?

I'm sure we can all agree that despite what happens in nineties comedy shows, butt slapping is unlikely to be a type of touch that we'd welcome in the workplace. In fact, many have criticized the show for making light of male experiences of unwanted touch. But what about other types of touch, like a handshake or a pat on the back? How confident would you feel about engaging with these touch types at work?

In 2011, a group of researchers led by Bryan Fuller of Louisiana Tech University published a study called "Exploring Touch as a Positive Workplace Behavior." It attempted to establish what factors make some individuals more likely than others to be open to engaging in workplace touch.[10]

To do this, over 400 working adults were surveyed. The researchers asked each person questions about their views on their ability to use touch positively in the workplace—what researchers call their "touch confidence." For instance, they asked how much participants agreed with statements like "I can use touch to form stronger working relationships with others."

The survey also asked about anxiety when engaging in touch at work. For instance, how much participants agreed with statements like "It scares me to think that I could damage my relationship with someone at work if I touch them and they take it the wrong way."

As you might expect, people's confidence in using touch at work and their anxiety about touch in the workplace were related. People who felt more confident about their ability to use touch positively were less anxious about touch in the workplace in general.

But the researchers did not simply look at this relationship alone. They also examined other individual differences contributing to confidence and anxiety about touch at work. The survey asked about a range of factors, such as how well participants thought they could achieve goals that they set themselves or perform tasks.

People who held stronger views about their general ability to perform tasks successfully and achieve personal goals showed more touch confidence at work. These people were also more optimistic and upbeat. In other words, their greater confidence was not just related to touch. To draw on the *Friends* example one more time, we might expect that a boss like Doug would not only feel confident slapping butts at work but also believe that he would perform well in other settings—sports, singing, learning new things, and so on.

Personality also predicted differences in touch confidence at work. You'll likely recall from earlier chapters that more extroverted people can feel more positive about everyday social touch. You might expect they would also have more confidence in using touch at work. You'd be correct. People who scored higher in extroversion showed higher scores in their confidence around using touch at work. Conversely, people who reported being shyer had more touch anxiety in the workplace.

In a nutshell, although Doug and Chandler might work in the same place, they can have very different preferences and thoughts about how to touch at work. Different features, such as personality, can affect perceptions about how confident we feel about engaging in touch in the workplace. And these are only the factors we know about. There are likely to be more (think of all the individual differences we've mentioned throughout this book), but so far, the science of workplace touch has not fully explored these.

POSITIVE, BUT NOT: TOUCH AND TRAIT IMPRESSIONS IN THE WORKPLACE

Before I move our discussion on, it would be remiss of me not to mention one final experiment that Bryan Fuller and colleagues conducted on the potential for touch to be a positive workplace behavior.

It saw them survey almost 235 pairs of supervisors and supervisees who worked together. To make that a bit easier on the tongue, we'll refer to them as managers and staff members from here on.

As in the original studies, managers completed measures of touch confidence and touch anxiety at work. They were asked how often they touched their staff members and their need for touch. For instance, how much they agreed with statements like "I consider myself a touchy-feely person."

Staff members were asked how often their manager touched them in a typical working shift. They were also asked about their own need for touch and their views of their manager. Some questions about their manager included how good they felt the manager's interpersonal skills were, whether they perceived the manager to be supportive, and how likeable they were.

On average, touching more frequently at work was linked to several positive impressions from the staff member towards the manager. These included perceptions of a manager's interpersonal influence, sincerity, likability, and support. Higher ratings in each of these were related to more frequent touch from the manager in the workplace.*

But does this mean that touch always leads to positive evaluations of the manager? No.

---

* A vital caveat to these findings is that we cannot be sure that it was solely touching that drove these effects. For instance, other behaviors that contributed to the positive evaluations may have occurred alongside the touch.

The trouble with touch is that you can rarely pin things down to what happens on average. Sure, touch was linked to some positive outcomes. But the extent to which manager touch led to high ratings of interpersonal influence depended, as you might expect, upon the individual staff member's need for touch.

People who reported needing more touch were more likely to rate a manager who often touched as being higher in interpersonal influence. In contrast, the relationship between the amount of manager touch and perceived interpersonal influence reduced when the staff member's need for touch was low.

To put that into a more applied situation: If a manager wants to ensure the social effectiveness of touch in the workplace, they must be acutely aware of the needs of the team member they are touching. Even a well-meant touch to convey positive support may not yield positive outcomes if it does not align with the needs of the staff member being touched. When making physical contact with others at work, one must be keenly aware that people differ in their need for touch. Think of the hearty mutual backslapping of some workplaces and how differently this might be received by someone who does not enjoy close contact with colleagues or finds touch overwhelming. This difference in our desire for touch can significantly determine the likelihood that touch will lead to a positive outcome in the workplace.

> **How can I get better at judging tactile preferences at work?**
>
> The science of touch may not know all the individual differences that contribute to variations in people's thoughts and feelings about touch at work. Still, there are lessons that we can draw from elsewhere to help navigate social interactions in the workplace. One example is cultivating emotional competence. Emotional competence relates to our ability to interpret and

respond to our own and other people's emotions. Research in this field teaches us that we can improve our workplace social interactions by increasing awareness of our own feelings, paying attention to nonverbal cues, and trying to see things from other people's perspectives. When you consider it, there is an overlap between these themes and some earlier takeaway messages from this book. To help our tactile interactions at work, we may want to try to improve our touch competence by:

1. Improving awareness of our thoughts and feelings about touch in workplace interactions. E.g., asking: Am I a touch-confident person? Would I mind if someone touched me at work?
2. Working hard to not automatically assume that our preferences map onto other people's. E.g., being aware of the false consensus effect we discussed in Chapter 7.
3. Paying attention to our own tactile behaviors towards others. E.g., asking: Do I touch people often at work? Would other people say that I often touch in the workplace?
4. Being respectful and empathetic towards other people's social signals, even if they differ from our own desires. E.g., asking: Is that person taking a step back as I approach them?
5. Learning about the different preferences of other people and different cultural norms towards touch at work. E.g., asking: If I saw my colleague being touched by someone else, how do I think they would feel?

SHAKE IT OFF

"I don't think we should ever shake hands again, to be honest with you,"[11] said Dr. Anthony Fauci in April 2020, around one month after COVID-19 was declared a pandemic. At the time, Fauci was the head of the US National Institute of Allergy and Infectious Disease. Others went further, with infectious disease expert Gregory Poland stating in one interview, "When you extend your hand, you're extending a bioweapon."

In the early stages of the COVID-19 pandemic, handshakes were a hot topic for discussion. The gesture was a historical standard for greeting and signaling agreement across a range of professional and personal contexts, but now many asked whether it had seen its day.[12]

At the time of writing, however, roughly two years (and many elbow bumps) after the start of the pandemic, it seems handshakes are back. They are present in sport, politics, and business meetings. Post COVID-19, you may also have seen or experienced the new awkward social interaction between people unsure whether to shake or elbow bump or do a combination of the two. As columnist Pilita Clark wrote in a 2021 article for the *Financial Times*, "Business meetings are back, bringing disastrous collisions between shakers, bumpers, and fist knockers . . . I can report that, without a doubt, it is bedlam out there."[13]

When you think about it, the idea that handshakes might have disappeared after the pandemic seems a bit like wishful thinking (at least if you are a shake hater). They've been around for centuries. In the Pergamon Museum in Berlin, a sculpture on a gravestone from the fifth century BCE shows two soldiers shaking hands. Despite pandemics and many other seismic events, there is a stubbornness to this tactile ritual that has kept handshaking with us for many years.

What contributes to making handshakes so ingrained in society? Some have argued that it is their meaning; historians have noted that handshakes have been depicted as a sign of trust and agreement in many ancient civilizations, including the ancient Greeks and Egyptians. Others have connected the broader use of the gesture as a greeting between friends with the seventeenth-century Quaker movement. Supposedly the Quakers broke from a convention of bowing to greet one another by giving a hand of friendship.[14]

In more modern history, we might want to consider how the handshake has often been an accompaniment to many notable moments, like when Nelson Mandela and the then South African president F. W. de Klerk shook hands in 1990 after announcing an agreement to talks about ending white-minority rule. I would argue that many poignant historical occasions would have landed very differently to observers without a handshake. Sure, people could have hugged, elbow-bumped, or not touched at all, but this just wouldn't have had the same impact. There is something meaningful and important in the physical contact that comes with the ritualistic handshake.

World leader or not, it is likely that you will have encountered handshaking at some time in your life. Handshakes are a prominent part of Western culture and increasingly creep into interactions worldwide. In the Touch Test, we found that the gesture people felt was most appropriate for interacting with their boss at a work social event, regardless of whether that boss was described as male or female, was a handshake.

Despite its prevalence, handshaking is a greeting that can leave many divided. Some love it; some hate it. Many find it a crucial part of etiquette; others find it gross.

The regularity in which we encounter handshakes can also lead us to overlook how impactful the simple exchange of grasping hands can be. Let's take one example: shaking hands before a job interview. It's a common activity. You might think it is culturally significant, but probably nothing more than a brief opening interaction between the candidate and their potential employer. You might suppose that the outcome of this initial physical exchange dissipates relatively quickly, but you would be wrong.

Research published in 2008 found that the perceived quality of a person's handshake was related to job interview success.[15] Close to 100 students participated in this study. Specially trained assessors rated each person who took part on their grip, strength, duration, vigor, and eye contact during a handshake.

In addition to their handshake assessment, each research participant attended a mock interview. Notably, the handshake assessors did not participate in the interview or share their feedback with anyone involved in it. This meant that both interviewees and interviewers had limited awareness of the handshake evaluations.

People who scored higher with the handshake assessors were the most hireable by the interviewers.

Let's unpack that a bit. The handshake assessors' views on having a firm grip and looking the other person in the eye during a brief handshake were predictive of the mock-interview outcome. Even after a 30-minute social interaction, where all sorts of other cues could influence how interviewers rated the candidates, the handshake mattered! It is incredible, and somewhat disturbing, that such a brief exchange can carry so much weight.

It doesn't just stop there. Handshakes have been shown to influence judgements of personality traits. They can impact our perceptions of the confidence and competence of others, and our decisions to engage in deals or cooperation with people we interact with.

An extensive demonstration of the latter comes from experiments led by researchers at the University of California, Berkeley, and Harvard Business School. The research, published in 2019, recorded whether pairs of participants did or did not shake hands before various social interactions.[16]

In one situation, the interacting partners engaged in negotiation outcomes—like negotiating the price of a car. Shaking hands was associated with a higher success of joint negotiation outcomes, regardless of whether the pairs knew each other previously. A simple shake brought people to common agreement more easily.

In another scenario, participants engaged in a mock-job-offer negotiation: deciding upon the salary, start date, and office location. Here, half of the participants were told that it was customary for people to shake hands before starting a negotiation. In contrast, the other half were told that it was customary to sit across from their partner when starting a negotiation.

Those who shook hands achieved higher joint outcomes and showed more cooperative behaviors. They lied less, leaned towards each other during the interaction more, talked for a bit longer after they had reached an agreement, and shook hands at the end of their discussion.

You might suppose that part of the reason handshaking played a role in promoting cooperation was because the experimenter told them it was expected. In follow-up experiments, the researchers changed their approach. Rather than telling both parties in a pair that handshaking was customary, they only told one. The other received no instructions, simply an outstretched hand from their partner. If the verbal instruction mattered, you might expect one partner to be more cooperative than the other.

Once again, handshaking increased cooperation and, importantly for both partners, contributed to collaborative intent irrespective of whether they were instructed to shake hands or not.

What was also fascinating in these follow-up experiments was that the negotiation context was more adversarial. In one scenario, people engaged in a real estate negotiation. Here, the buyer had insights into additional information that the seller did not. The buyer was aware that the laws would soon change, allowing them to develop the land to make it more valuable than it currently was. Not disclosing or lying about this information was in the buyer's interest.

But the buyers who shook hands with the sellers were less likely to do this. They were more truthful. This meant that the negotiation outcomes were more equal, and the sellers did better. Even when cooperating meant gaining less in the exchange, the handshake increased willingness to work together for the common good. The next time I'm selling my house, I'll be sure to shake the hands of the people who come to view it.

What all these findings show is the surprising power of a simple handshake. Not only can it lead to more cooperation, it can also even do this at one party's expense. Handshakes appear to be more than a ritualistic behavior. The physical contact goes much further, signalling cooperative intent. A handshake is a simple gesture with a powerful effect.

To touch or not to touch

To conclude this chapter, let's return to where we began: a scenario where a manager greets people in the office with a pat on the back and a rub on the shoulder. Our question was: Should these types of tactile interactions occur in the workplace? The unsatisfactory answer to this question is that it will likely depend on several factors.

As we've seen, touch can carry many benefits for building trust, perceptions of gratitude, and support. But it can also carry risks. These can come from extreme examples of inappropriate touching to tactile interactions that may not be as egregious but can still make a staff member feel uncomfortable. Even everyday interactions often deemed appropriate, like handshaking, can provide good and bad outcomes—signalling agreement and cooperative intent but also risking bias or negative impressions.

Putting all this together, we're left with a bit of a dilemma: Touch carries risk and, for many, is inappropriate at work. But it also benefits team performance and interpersonal bonds that may be helpful between colleagues. What is the best way to approach the murky world of touch in the workplace? If I'm honest, the science isn't clear.

When evaluating whether to touch in the workplace, every individual must weigh many factors. A major one is our colleagues' desire and consent to be touched. Another is what the workplace norm around touch is. These norms will be set by the collective group of employees and by workplace policies, which will have to carefully balance the risks and benefits of touch for the collective good of a community.

The difficulty for workplace policymakers is that evidence about touch in work settings is sadly in short supply. There's an urgent need for collaboration between researchers and employers to understand touch in the workplace better. This will help support a culture of evidence-informed practice that can help employers take the best but lose the worst bits of workplace touch.

For now, perhaps the safest bet is to establish a norm that works for the majority within the organization. This norm has an important exception—that everyone is asked about their touch preferences. If you work in an environment where touching is deemed appropriate, you should be able to opt out.

CHAPTER 10

Digital Touch: The Future of Touch in Our Society

Picture an all-black, tight-fitting full-body synthetic suit, paneled across the torso, exaggerating the chest and abdominal muscles much like the fabric armor worn by the late Chadwick Boseman in Marvel's *Black Panther*. This suit has built-in motion-capture sensors and biometrics to record the wearer's heart rate, skin conductance, and other key health indicators. It has electrostimulators providing tactile sensations all over your body. It is designed to act as a human-to-digital interface.

The suit I am describing might sound like something you'd find in a science fiction movie, a suit that can enable the wearer to explore impossible virtual worlds—to escape the boundaries of their existence through full-body tactile immersion.

Believe it or not, it exists. It is known as the Teslasuit (no relation to the electric car or Elon Musk).[1] Advancements in tactile technology—also called haptics—have helped develop a physical suit that can provide tactile feedback in virtual worlds. The suit enables the sharing of tactile sensations without physically making contact with another person or object.

To give just one example of this in action, we can look to the telecommunication company Vodafone's demonstration of using 5G technology and the Teslasuit to share touch between people close to 100 miles apart. In this example, two rugby players were training at separate stadiums in the UK—one in Coventry and one in London.

One player tackled a pressure-sensitive tackle cylinder—a bit like a big punching bag. The cylinder detected whether the impact of the tackle was hard, medium, or light. This information was sent via 5G to the software controlling the Teslasuit. Almost 100 miles away, the other player felt the experience through the suit. Digital touch helped them instantly experience the force of the tackle in specific muscles on their body. Two players—miles apart—can train and share tactile experiences together.

Although sports and virtual-reality gaming are clear applications for the Teslasuit, the potential of this technology makes the mind boggle. In a 2021 presentation, the company behind the suit provided many examples of its use. Working with space agencies, the Teslasuit has been used in virtual-reality-based research trials investigating space travel's effects on physiology and psychology. By making virtual reality feel more natural, these organizations have investigated how our body responds when we feel like we are in environments with variable gravity.

Back on "virtual Earth," haptic feedback and biometrics provided by the Teslasuit have been used to train workers responding to emergency scenarios. This includes on oil rigs and in defense settings. The company reports that tactile virtual-reality training is underway with police officers, firefighters, and even pilots.

I also could not help imagining another situation where devices like the Teslasuit might be valuable in the future: to help people connect with each other. Just think of those who report falling short of social connections with others in their lives. Could virtual reality and tactile suits someday help to improve long-distance relationships so that people can share sensations with those they cannot be physically close to? To help answer this, perhaps we should start by exploring the various ways touch tech is being developed for use around the world.

THE FUTURE OF WORK

Undoubtedly the 2020s have seen seismic shifts in how people work. Before the COVID-19 pandemic, Americans were reported to spend 5 percent of their working time at home. In 2020, this was estimated to have risen to 60 percent, as organizations rushed to ensure work could continue when staff were unable to be in the office.[2]

Many people now say that they would like to work from home more than before the COVID-19 pandemic: One survey of thousands of Americans suggested that the average employee would like to work from home close to half the time.

Trends like this led Mark Zuckerberg, Meta's CEO, to conclude that "we should be teleporting, not transporting ourselves" to work.[3] Working where we want to be based while being "present" in working environments.

Working more productively. Working more flexibly. In a way, that could reduce our carbon emissions and, in doing so, help the planet.

The pandemic certainly opened minds to the potential of remote working. Some companies responded to this by introducing mobile offices. Others, like Meta, have been exploring the possibility of augmented- and virtual-reality technologies to take remote working to the next level. To help people connect and collaborate in online workspaces that feel real, while still preserving the benefits of working from home.

Meta's plans for a redesigned future of work are bold.[4] The company intends to use technology that enables the person working in the virtual environment to switch between real and virtual worlds at any time, with a simple flick of a button powered through their virtual-reality headset. This would give people the best of their home office environment and a virtual world where they can collaborate and authentically interact with coworkers.

Imagine no more staring at the dirty dishes while working from your kitchen table. Instead, put your VR goggles on and find yourself transported to a shared exotic office overlooking a tropical lagoon, all while wearing your comfy loungewear and slippers.

If you think this sounds leaps and bounds more advanced than the freezing video calls and weak Wi-Fi connections that most of us experience while working from home, you're not alone. For many, the early stages of remote working during the COVID-19 pandemic were often best described as survival mode.

The initial novelty in my workplace was quite enjoyable, but it soon wore off. Establishing new relationships with colleagues in our tiny boxes on a screen felt tough. It felt harder to collaborate. After several months of online-only working, the words "let's jump on a quick call" can still send a shiver down the spine.

Yet what Meta proposes pushes the boundaries of remote working to the next level. The future of home working is no longer a question of whether people can get online to work remotely. It is a question of how remote working can begin to feel genuinely engaging. It is about feelings of authentic and lifelike sensations in virtual environments.

But how do we get to this? What will help companies like Meta to create a genuinely authentic virtual workplace?

A critical aspect of the future of virtual work is the power to mimic our ability to share and experience touch.

For virtual reality to be immersive, we require sensations that match our lived experiences and expectations about our sensory world. If we reach out for someone, we expect and want to feel the sensation of another person. If we manipulate everyday objects, we want to be able to feel something in our hands.

Consider the sensations connected to feeling a virtual keyboard under your fingertips while typing. The feeling of a colleague giving you a virtual high-five. Or even how you experience playing virtual table tennis with your office mate. The ability to truly harness the potential of virtual reality in the future of work is closely connected to how the virtual world will feel.

This isn't just true in workplace virtual reality. As virtual reality becomes increasingly part of our actual reality, the use of tactile technology will be important in the fields of virtual entertainment—like gaming and sharing experiences at virtual music concerts.[5]

There are now companies that are trying to make people experience touch in entertainment environments without any devices attached to the body at all. Ultraleap is an exciting young company based in Silicon Valley in the US and Bristol in the UK. They combine advanced motion tracking and unique haptic technology to give the sensation of touch in midair,[6] creating feelings of touch via speakers that emit ultrasound waves.

I once got to experience this technology and was fascinated by the experience. Nothing touched me, but I felt ripples of touch moving across my hands by hovering them above the speakers.

Aside from being very cool, ultrasound technology like this has several compelling uses, from providing haptic cinematic and gaming experiences without needing physical devices to be attached to the body to touchless technology solutions that may help build new self-service interfaces.

Much of this promise hinges on a simple assumption: that we can understand what midair touch actually means. Think about it for a moment. Imagine that you are playing a video game, and one of the characters grabs your arm to express surprise. You feel this through midair ultrasound touch. I expect this will be a somewhat immersive experience. But it would be even more immersive if you could understand what the tactile experience was trying to convey.

We learned earlier that emotions can be conveyed by touch in the real world, but can we understand emotions expressed by midair touch? Remarkably, the answer seems to be yes! Researchers from the University of Sussex explored the communication of emotions through tactile stimulation in midair.[7] One group of people was shown pictures of emotions and asked to generate emotions corresponding to them with midair touch. A different group of people was later asked to interpret the emotions from midair touch alone. This second group was able to identify the emotions expressed. Much like everyday touch, people could potentially use midair touch to successfully convey emotions too!

THE TERMINATOR

Using touch technologies to support future innovations is not just a priority for organizations exploring the potential of virtual reality for office working or entertainment. Such technologies can also be seen in more everyday settings, such as introducing vibrations to your car steering wheel to warn you of an approaching hazard. Or using tactile feedback to a machine operator to let them know they are about to engage in unsafe behavior.

Technology is increasingly harnessing our sense of touch to help build safer interactions. Take surgical robots as an example. These can be used to conduct operations on patients under the control of a trained operator from anywhere in the world. A revelation in global healthcare, they could make a real difference in the accessibility of healthcare in the future by reducing surgical times and helping with the precision of treatment.

While the technology is full of promise, I suspect that if I were a patient undergoing surgery by a robot being controlled by someone hundreds of miles away, I'd like to know that my surgeon could feel what the robot was doing to my body. I'd want to be confident that they could conduct the surgery safely and accurately. An essential factor in achieving this is tactile feedback: Studies show that giving an operator haptic feedback when controlling these robots can lead to better accuracy and reduced operating times.[8]

A very different profession where haptic robots are increasingly used is bomb disposal. Here, the aim is increased safety by using robots to help bomb disposal teams deal with dangerous explosives from a distance.

These robots are impressive and expensive pieces of technology. Imagine that Pixar's WALL-E had opted for military fashion, complete with high-definition cameras, adjustable arms, and tank-like all-terrain treads.

Notably, modern bomb disposal robots have sensors to support tactile feedback.[9] This feedback is designed to help recreate tactile experiences for the operator when dealing with dangerous explosives. It sends the operator vibrations when the robot touches elements of a bomb, helping to guide bomb disposal teams in their challenging and often high-risk work.

Part of the reason for this is because when we manipulate objects and use the dexterity of our own hands, we rely on touch. This means that if we are to fully take advantage of robots becoming more lifelike and dextrous, there is an increasing need for tactile feedback to be shared with the human operator.

*

The need to support greater control and understanding of sensations experienced by robots has seen the development of other forms of touch tech, for instance, haptic gloves designed to give people control of advanced robotics at their fingertips.

In 2020, Converge Robotics Group debuted their Tactile Telerobot, a fantastic piece of technology that is helping to take operator–robot touch to a new level. Described by some as "perhaps the most complex robotic hand on earth" and looking "a bit like the Terminator's hand when he rips off his skin,"[10] this piece of technology appears to have come straight out of a science fiction movie.

By using a glove that provides tactile feedback through vibrations, the operator of the Tactile Telerobot can move robotic fingertips with incredible dexterity. In company promotions for the glove, you'll find videos of people using their robotic fingers to pick up items, make gestures, and even try to complete a Rubik's Cube.

Although this technology is at an early stage, it offers the potential to enable people to handle objects and complete behaviors with tactile feedback across vast distances like never before. The advances in dexterity and touch-based feedback from the Tactile Telerobot look game-changing. It could offer huge potential to aid teams working in dangerous situations, medical settings, and beyond.

THE HUG BOT

It's easy to think of robots at work as something futuristic, but they are already a part of our present. You can now find robots in active use in many industries—healthcare, delivery, and public-transport robots are a few examples. As they become an increasing part of our world, this raises an important question: How do we feel about interacting with robots in society?

There are several ways to interpret this question. Many have discussed fears about robots replacing human workers and what this may mean for living a meaningful and happy life. Others point out that if a robot took over repetitive tasks, it might free up more time for people to interact.

Some also go as far as to suggest that if we designed robots in such a way that they were true collaborative colleagues, they might provide a source of social interaction themselves.[11]

How could robot designers facilitate this? One important feature will be how robots touch.

In 2009, researchers from the University of Amsterdam found that people rated a humanoid robot as more reliable and less machine-like when interactions with it included touch. In the study, participants watched videos showing a robot offering to help a person. Notably, the robot touched the person in some of the videos, but in others, it did not. The participants favored the interaction with touch; they preferred robots to touch when offering help.[12]

Other researchers have shown that some of our preferences towards human touch extend to touch from robots. In 2019, scientists from the Max Planck Institute for Intelligent Systems and the University of Pennsylvania conducted an experiment involving a robot that gave hugs for different durations. They also varied the amount of pressure. Short hugs, long hugs, firm hugs, and softer hugs, all from a robot.

Short hugs lasted for one second. They contrasted with an "immediate-release" hug, which saw the robot hug the participant until the person being hugged released their arms from the robot's back. Another hug lasted an extra five seconds after the person being hugged let go—a "too-long" hug.

After each hug, the person who was hugged was asked to rate the robot on various dimensions, such as how caring or comforting it was. The results showed that following immediate release and the too-long hugs,* the robot was judged to be more social, caring, happy, and comforting. Like hugs between people, duration mattered, with too-short hugs being viewed less favorably.[13]

These findings have implications for how robots that touch might be designed and used moving forward. Increasingly we will see robots interacting more with people, especially in healthcare settings. If we want to maximize their potential to support health

---

* We should remember that the too-long hugs lasted only five seconds too long; imagine if a robot kept holding onto you for minutes rather than seconds.

and well-being, we need to know how to optimize their tactile interactions with humans.

The results on hugging robots also had me wondering just how far robot hugs could go. What might an optimized human–robot interaction look like?

If I had a hugging robot, I think I'd like it to understand that most people will want a hug for a few seconds, and for it to start by hugging me for that amount of time. But to get the interaction right, I think I'd like my hug bot to be a little smarter than that. Perhaps it could learn that I prefer a hug that lasts four seconds exactly, and optimize our hug for that duration.

Taking that a step further, a supportive robot could also detect if I hugged it for longer one day, and possibly check that I'm doing okay.

You may think this is far-fetched, but a few years ago, some scientists were developing a personal assistant robot that could learn to adapt its behavior based on the tone of voice of the person interacting with it.[14] This type of robotic learning combined with robots that feel could go a long way in allowing people to share feelings of social support from robotic touch. It would be a dramatic innovation, especially when you consider the high proportion of people who report a lack of touch or affection in their lives.

While we wait on these developments, perhaps I'll have to settle for HuggieBot,[15] a robot that stands there, arms open, waiting for you to hug it. When it feels you close your arms around its back, it hugs you. When you release your grasp, it lets go.

THE PET THAT DOESN'T POOP

Research on hugging robots shows us that how robots touch us matters, but what about how we touch them? Recently, I was given a tour of Chelsea and Westminster Hospital in London as part of a new research study I'm undertaking.

For those who do not know, Chelsea and Westminster Hospital is an amazing place! It's unlike any hospital I've visited before. Part of the reason for this is that it was developed with the arts and a sensory connection to the environment in mind.[16]

There are over 2,000 artworks in the hospital's collection, including some by world-renowned artists Tracey Emin and Julian Opie. All are carefully exhibited, with the addition of indoor gardens and cinemas to bring the hospital to life.

My tour did not just take in the fantastic design of the hospital. It also enabled me to hear firsthand about some of the great new technologies that staff are working with to support patients in healthcare settings. One that stuck with me was the robotic companion pets: a dog-like pet called Miro[17] and a seal-like pet called Paro.[18]

Remarkably, research has shown that when people interact with these types of robotic companions, they can experience reductions in anxiety, blood pressure, and pain. This pattern of results has tended to be found when focusing on specific vulnerable groups, such as older people in residential care.

For instance, one study led by researchers from the University of Auckland compared the blood pressure of elderly residents before, during, and after interacting with Paro. Paro is modelled on a Canadian harp seal and is covered in white fur.* People in the study would stroke and touch it, and Paro responded by moving and making the noises of a baby seal. You might even say it acted as a virtual pet would.

The results of the experiment were fantastic. The elderly residents were more relaxed. They also had lower blood pressure

---

* I was told that a harp seal was used instead of a dog to avoid people getting distracted by the likeness of the robot to an animal that they may have met in the real world. I suspect most of you will have had fewer encounters with a real-life harp seal than a real-life dog.

when interacting with the seal.[19]

We know that animal companions can be good for people's stress and well-being. Robot seals might provide an elegant solution to bring those types of benefits to people who cannot have furry companions in their lives.

It also turns out that robot companions don't necessarily need to be furry to bring benefits to wellness. Even engaging with plastic robot dogs can reduce loneliness in elderly participants living in nursing homes. Researchers from Saint Louis University School of Medicine examined the effects of interacting with a living or robotic dog on levels of loneliness. The robotic dog in this study was an Aibo robotic dog,[20] not to be confused with the Miro dog that I encountered at Chelsea and Westminster Hospital: Robot dogs come in different breeds too!

Both the living dog and the robot dog contributed to improvements in older adults' levels of loneliness.[21] Remarkably, there was no statistical difference in loneliness improvement between those who interacted with the real dog and those who spent time with the robotic dog. It's a finding that I'm feeling strangely offended by, though that might be out of loyalty to the family dog, Loki, who is snoozing by my feet as I write this.

Putting Loki's feelings aside, we know that having pets can benefit well-being and loneliness. Access to a real-life pet is not always possible, especially for elderly nursing-home residents. These findings offer an intriguing possibility for alleviating loneliness in older people.

They also beg the question: If robot touch can help the elderly, can it help other groups too? You might recall from Chapter 4 that alongside the elderly, another group that is often at high risk for feelings of loneliness and mental health difficulties is younger adults. Do younger adults benefit from interacting with robot pets?

Researchers from the Ben-Gurion University of the Negev recently asked this very question. In a study of a group of young adults, they assessed pain and happiness before touching a robot seal, during touch, and without touch.

Touching the seal led to a larger decrease in pain ratings than no touch. There was also an increase in happiness ratings compared to the baseline following human–robot touch. Just like for older adults in nursing homes, robot touch benefits young adults,[22] once again demonstrating that for those who can't have a pet or simply don't want the responsibility, robots can offer benefits too!

Whether in healthcare settings or other day-to-day interactions, robots will likely become an increasingly present part of our world. This increase will naturally expand the possibility of human–robot interactions through touch. The emerging field examining the impact of companion robots is beginning to build a convincing picture that there is potential for human–robot social touch to support wellness moving forward.

USING TOUCH TO NAVIGATE THE WORLD

The future of touch technology is not all about expensive pieces of robotics. Other, more subtle forms of tactile technology exist. Many of these relate to a growing focus on inclusive design: building products and services that remove barriers faced by people with different levels of capabilities.

We saw earlier how companies are exploring increasingly vivid and rich virtual worlds that combine vision, sound, and touch. An inclusive design approach would ask: Can everyone use this service? How would someone who is deaf experience this virtual world? How would someone who is blind navigate their way through a complicated virtual environment?

The use of touch tech to foster a greater sense of immersion in digital environments is not only critical for sighted individuals. Tactile feedback is also central to building a greater sense of immersion and agency in virtual and augmented-reality environments for people with sensory difficulties.

One recent advancement is a novel white cane controller that uses haptic feedback to help blind or low-vision virtual-reality users navigate through virtual environments. Researchers from Microsoft indicate that the cane controller has a lightweight brake mechanism, contact vibrations, and spatialized audio. Using a scavenger hunt game, the researchers showed that the cane enabled blind users to navigate a complex virtual environment and locate targets while avoiding collisions.[23]

Another powerful example of engaging with haptics to support a more accessible world comes from the science of sensory substitution. Sensory substitution techniques attempt to convey experiences from one sense by representing them with an alternative sense. They have often focused on converting visual or auditory signals into touch, for instance, a sound felt as a vibration on the skin. Or transmitting text or graphical information via tactile electronic displays.

To give you an idea, we can turn to the work of David Eagleman, a neuroscientist, author, and entrepreneur. Some years ago, his research team asked: Can information from one sense be converted and understood through another sense? They developed a vest with vibratory motors stitched into it, which captured sounds in the environment and algorithmically converted them to tactile sensations on the skin that varied according to the sounds detected.[24]

In a study published in 2021, Eagleman's team converted their vest technology to work within a watch that translates sound into vibrations. They found that participants wearing the watch could identify up to 95 percent of the environmental sounds tested in the study through these vibrations (the average was 70 percent).[25]

It was also found that practice improved performance. Deaf and hard-of-hearing participants who used the watch regularly over a month got better at perceiving sounds converted to touch. The researchers concluded that the watch offers a new wearable technology to interpret auditory stimuli for the deaf and hard of hearing.

These findings have striking importance. We live in a world where the number of people with sensory difficulties is expected to rise. In this world, the development of touch-based technologies like that of Eagleman's team offers rich promise in using the sense of touch to help dismantle barriers to hearing.

If Eagleman's watch had me excited about the future of touch technology to aid accessibility, then what came next had my brain doing somersaults. Believe it or not, there are people in the world who are building artificial limbs that send signals to the brain to share the sensations they feel.

Artificial limbs (prosthetics) play a vital role in supporting mobility and independence in daily life for many people with limb loss. Their use has been traced back to as early as the ancient Egyptians. Some 5,000 years later, the modern generation of prosthetic limbs has a new twist, combining electronics, hydraulics, and tactile technologies to improve their functionality.

In 2019, a proof-of-concept study led by scientists from the Swiss Federal Institute of Technology Zürich found that adding tactile and motion feedback to a prosthetic leg had significant benefits.[26]

Signals from sensors were sent to the brain via neural implants, resulting in better walking speed and self-reported confidence from people wearing the artificial leg. They also had less fatigue when they received this feedback than when they had no feedback. In other words, adding tactile sensory information to artificial limbs helped the functionality of the prosthetic limb.

The potential of research investigating interactions between touch and artificial limbs is exciting. Any accessible developments that help to make prosthetics more usable have considerable promise. We must be mindful that this work is still relatively early in development. Nevertheless, as we move forward, we can expect more from this emerging field.[27]

THE POWER OF HAPTIC WELLNESS

Robots, virtual-reality suits, midair touch: The future of touch technologies has vast applications. But the area that I am perhaps most excited about is the emerging use of haptics in wellness. In 2020, Digital Catapult—a UK agency promoting digital innovation—published a report that suggested haptic wellness technology will be one of the fastest-growing sectors over the next few years.[28]

Haptic wellness is all about merging advances in wearable devices and advances in touch tech to support the growing focus on improving the quality of life in society. Wearables will not only collect data but also provide real-time feedback to the person wearing the technology through touch.

Some examples emerging from this space include wearables that aim to reduce stress or anxiety by providing silent but soothing vibrations. These products aim to nudge our heart rate and other aspects of physiology to states that might help our body adapt and recover from stress and anxiety. They tend to do this by matching these vibrations with rhythmic heart rate patterns associated with changes in activity within the nervous system.

Take Doppel as an example. Doppel is a wearable device that delivers heartbeat-like tactile stimulation on the wrist.[29] In 2017, researchers from Royal Holloway, University of London, found that using the device reduced physiological arousal and self-reported feelings of anxiety.

In the research, two groups of participants wore the device before a socially stressful situation—the anticipation of delivering a public speech. Doppel was turned on for one group to produce a slow heartbeat-like vibration. The other group acted as a control group: They did not experience these slow vibrations. Participants in the vibration group showed lower physiological arousal and anxiety levels than the control group. These are fascinating results that illustrate how something as simple as vibration can be used effectively to support wellness.

Even more astonishingly, we are beginning to see ways in which the next generation of wearables may not even involve anything visible at all. When we think of wearables, many will likely first consider smartwatches, glasses, or jewelry. Few of us would mention a new technology: electronic skin, or e-skin.

E-skin is best described as an electronic system that aims to imitate features of human skin. It is thin, lightweight, durable, and full of sensory properties. Some e-skin can deform like natural skin, and even self-heal after being torn or cut. Like real skin, it can also fuse back together and stay stretchy.[30]

The applications of e-skin are vast. Robot designers are using its unique properties to help robots touch. To give just one example, some e-skin can sense force. This could help with those robot hugs we spoke about earlier.

It can also be more fine-grained than that. Researchers have used force-detectable e-skin to help robots gently or firmly touch different objects without damaging them, such as a raspberry or a Ping-Pong ball. This would be a helpful addition if you want to use a robot in the kitchen to help you make dinner, or in surgical settings.

It is not simply robots where e-skin is proving useful. People can use e-skin as a wearable on the body and inside it too. Some imagine a future where wearable e-skins could help more routine remote healthcare.

Many of you, like me, may have been forced into the post-pandemic online doctor's appointment. With increasing pressure on doctors, and in some cases, difficulties in people getting to doctors' offices, there is a good chance that this type of consultation will become more common in the future.

While these appointments can be convenient, I must admit that sometimes I worry that things might get missed because my doctor can't touch or feel me.

Imagine a future where wearables placed directly on my skin could send relevant data to my doctor to help with these remote consultations.

Letting that imagination run wilder, imagine if you could feel your doctor touch you during your online consultation. As they move their finger across the image on the screen, you feel it on your body during the call.

This might sound fanciful. Yet it's not as ridiculous as you might think. In 2019, researchers from the City University of Hong Kong and Northwestern University developed a wearable interface that can share tactile experiences directly to the wearer's skin.[31] Reportedly they are now exploring how tactile experiences can be shared between someone touching a screen and the person wearing the interface.

Other use cases for e-skin extend beyond appointments with the doctor. Xenoma is an innovative apparel company.[32] They use e-skin in clothing to support well-being and health monitoring. One of their best-known products is their e-skin sleep and loungewear: pajamas that can monitor physiological signals from the wearer and environmental factors.

The pajamas can measure body temperature. If things aren't quite optimal, they can send a signal to smart-home features to tweak the thermostat. Reportedly, they can also send alerts if the wearer suffers a fall, which is particularly useful for individuals like fall-prone older adults.

Products like those from Xenoma create a reality where our clothes serve even more functions than they currently do. Smart clothing that can connect with other technologies is rapidly becoming a genuine part of our world.

This brings me to a final point: There is more to our clothing than merely its use to cover us. It can often be a sign of changes in society. Just think of the rise of the miniskirt and the colorful patterned print clothing in the 1960s; the shoulder pads, power suits, and leg warmers of the 1980s; or the ripped and faded jeans that were a staple of the 1990s grunge scene (a personal favorite of mine).

The fashion world has been changing in recent years. Our clothing is adjusting to our new post-pandemic world. Increasingly clothes aren't just being made for the physical world but also the digital.

An example of this is hugging shirts: shirts designed for people to send hugs to one another from afar. One company has been developing their HugShirt since the early 2000s.[33] It connects to your phone via Bluetooth; then, using a mobile phone app, you can record a hug and send it to another person wearing a HugShirt. Anyone can send these virtual hugs: The person sending it doesn't have to have a shirt themselves; they can record a hug on the app and send it anyway.

As a tool to help maintain human connection at a distance, the appeal of clothing like this is clear. A long-distance couple could embrace from afar; a parent could hug their child goodnight even if they couldn't be in the same room. For those with family and friends around the world, the opportunity to send a hug to loved ones is likely to have some appeal.

Of course, some ethical considerations come to mind here. What if someone were running late on a business trip one night and asked their employee to send a virtual hug to their partner or child on their behalf? I'm not sure how I would feel if I was on the receiving end of that hug.[34]

Nevertheless, when used appropriately, virtual hugs could be an elegant solution for people missing touch from a loved one when away.

There are also interesting questions about how virtual hugging might or might not mimic some of the results of actual hugging. Earlier we discussed how physical hugs can positively impact health and well-being. One reason for this is the social support that hugging implies. Could a hugging shirt provide this social support too? We do not know the answers to these questions, but I know that researchers like myself are keen to find out.

In an increasingly digital world, our clothing will likely advance with us, reflecting society's aspirations, realities, and values. In the context of touch, that may well mean garments that allow us to record and share tactile experiences virtually. There is great potential to widen our ability to share touch like never before.

Where do we go from here?

Before we close this section of the book, the final question I'd like to pose to you is: Where do we go from here?

After the COVID-19 pandemic took hold, one of the most common questions I was asked was: What is the future of touch? Will we shy away from hugging and handshaking? Will we be scared to return to the touch we were used to? As the world changes before our eyes, how will touch move with it?

An interesting thing about a major crisis like the COVID-19 pandemic is that people often believe it will be disruptive to trends that existed before it. Yet as Mauro Guillén notes in his excellent book *2030*, contrary to this conventional wisdom, we might expect that an event like the pandemic will "most likely intensify and accelerate—rather than derail" trends that existed before it.[35]

This means that the future of touch is likely about the trends present among us today. It's present in the work that we've just heard about from companies and organizations around the globe exploring the potential of touch technologies to reshape how we interact with both the real and the virtual world. The incredible technologies in development will bring digital touch more and more into our lives as we move forward.

In all these technological advancements, we might wonder if the way we touch will become vastly different. Will digital touch change our tactile world for the better, or will it replace human contact?

My view is somewhat equivocal. Do I think technology will help us to touch? Yes. Do I think trends in haptic wellness and wearables will drive exciting developments in health and well-being? Also yes. Do I hope this will make touch more accessible? Absolutely.

But do I think technology will replace human-to-human touch? On balance, I do not. Touch tethers us to each other. It's fundamental to the connections and bonds at the heart of our social world. And as we have seen throughout this book, it is something that many people in society crave; think of the number of people who report wanting more touch or affection in their lives. The future of touch offers hope for exciting and unique technologies that may help us with this. Nevertheless, we should expect that it will also be full of good old-fashioned human-to-human contact.

CONCLUSION

We started this book with a simple question: What does touch mean to you? Do you recall the words that came to your mind? I wonder if your thoughts have changed after learning about the science of touch.

You might remember the three words that nearly 40,000 people worldwide provided in 2020: comforting, warm, love.

Needless to say, a lot has changed since my team asked this question in early 2020. We have moved from the first few days of what became a global pandemic to national lockdown restrictions involving the prohibition of touch, mutations of the COVID-19 virus, and ongoing changes in societal norms to help the fight against it. Even more may have changed by the time this book is published.

There is no denying that the COVID-19 pandemic put touch in the spotlight. We became more aware that we can pass on viruses through touch. Many people grew more conscious of how they touched themselves and other people. Yet, despite COVID-19, when I ask this question to audiences in talks today, I regularly get a similar response: comfort, connection, care. These are the words that people tend to choose when describing what touch means to them most of the time.

Throughout our journey in *Touch Matters*, my aim has been to share the meaning and diverse nature of touch in today's world, and the science behind a sense that makes us human. I hope it has helped you understand what touch means, how it defines us, and how it contributes to health and well-being. If nothing else, I hope it has opened your mind to think about a sense that we often don't stop to consider. To make you a little bit more aware of the richness of your tactile world. To see touch for the good as well as the bad it can bring.

To close the book, I want to give you four key takeaways and reflections that have stood out to me as I was writing.

1. There is a thin line between good and bad touch, but overall supportive touch seems to improve our lives: When used well, touch can have positive effects on mental and physical health, development, and performance. Still, people can be skeptical about touching each other. One solution to this is to simply *not* touch. I would argue that this introduces problems of its own—think back to the adverse outcomes that touch deprivation can bring. Instead, the question we face is how to get more supportive (good) touch into people's lives. It is a challenge for us all. We can start to address it by first understanding our own touch desires and experiences. We can extend this self-reflection by putting time and energy into learning and discussing touch with others in an empathetic way. If we embrace touch with a compassionate and open mindset, it will help us seize the positive opportunities that supportive touch can bring.

2. Touch is something that people increasingly appear to crave, but it is a need many are falling short of receiving: More than half of a world sample felt they had too little touch in their lives in 2020. Three-quarters of Americans surveyed felt that their country was in a state of affection hunger when asked in 2015.

Those are startling numbers and hint at a worrying trend in society. As we've seen throughout this book, touch and affection are more than simple sensory acts; they carry emotional significance that can be important for happiness and health. In my opinion, the fact that so many people feel they are falling short of meeting their needs for affectionate touch speaks to a need for urgent action. There is a broader requirement for society to rethink how we can support people to interact and form meaningful relationships with others. If you do ever find yourself in a situation craving touch or affection, I have two messages for you: First, you are not alone; and second, there are options to support you. In this book, I've tried to give a few examples of strategies people might wish to consider to bring more touch and affection into their lives. I also recommend Kory Floyd's book *The Loneliness Cure*, which contains a variety of detailed strategies to support feelings of loneliness and affection deprivation.

3. There are many nuances to navigate in our tactile interactions; this creates opportunities as well as confusion. One thing that changed when writing this book was that I became increasingly aware of the need to help answer some of those everyday sources of social confusion and difficulty. For instance, what do I do when someone I don't know tries to hug me or if I have an overly tactile boss? Research has built a convincing picture that our touch preferences can differ for various reasons (e.g., personality, attachment, neurodiversity). Still, it has been less good at answering the question of how we can improve our ability to interact with each other through touch. This is a challenge but also an opportunity for us all. The fact that we have diversity in our lives should be celebrated rather than avoided. Let's take the time to understand those differences better and engage in deeper conversations about how we can relate to each other

based on our different preferences towards touch. It may open a door for us to form stronger connections.

4. The future of touch is bright. Without sounding too much like an advertising slogan, my final take-home reflection is to reiterate some of my conclusions at the end of Chapter 10. We are in an age where we will likely see rapid developments in tactile technologies, which, in my view, will dramatically shift our world in new directions. These could be robots that hug or artificial pets that can be stroked. They could be clothes that let you connect with loved ones from a distance. I genuinely hope some of these technologies will provide ways to make touch more accessible and helpful to people. I also have hope for how people choose to interact with each other physically through touch. For many, the COVID-19 pandemic stole touch from their lives. Sometimes it is not until you lose something that you realize how much you miss it. I hope we can try to nurture and maintain those positive relationships with touch that we all need.

APPENDIX: FURTHER MATERIALS

In addition to the quiz and persona tasks in Chapter 6, my research team has developed other measures that help examine individual differences in thoughts and experiences about touch in various contexts. For instance, how people experience touch in treatment settings, measures of mirror-touch synesthesia, and ASMR. These have been published in scientific journals and can be accessed at www.banissy.com.

A copy of the Touch Test, including access to the entire data set, is available here: https://reshare.ukdataservice.ac.uk/854471.

For those interested in learning about individual affection differences that extend beyond touch, I highly recommend looking at Kory Floyd's work: https://www.koryfloyd.com. On this site, you can also find measures of trait affection, affection deprivation, and affectionate communication: https://www.koryfloyd.com/research.

Keep an eye out for updates on touch, well-being, and other research via my social media (@mbanissy on Twitter and Instagram).

ACKNOWLEDGMENTS

This book is the product of many supportive people over the years. I began my journey in the field of psychology nearly two decades ago and have benefited from many mentors and colleagues during this time. Thanks to all who have contributed to intellectually stimulating environments during my time at the University of Hertfordshire, University College London, Goldsmiths, and the University of Bristol.

I have also been fortunate to work with my fantastic research team and collaborators. Special thanks connected to our work on touch go to Natalie Bowling, Katerina Vafeiadou, and Anna Lena Düren.

The work on the Touch Test would not have been possible without the fantastic support of the Wellcome Collection and BBC Radio 4. I am particularly grateful to Geraldine Fitzgerald, Claudia Hammond, and Chris Hassan. I am also most thankful to the thousands of people who participated in the research and those who shared stories with me throughout the broadcasts connected to the Touch Test. I am indebted to the variety of people who have contributed to this book through interviews and quotes. Thank you to those people who gave me their time and shared their insights and lived experiences of touch with me.

Belief in this project and months of effort behind the scenes led to the book you hold in your hands. I was deeply fortunate to be guided by editors Cara Bedick, Ru Merritt, and Pippa Wright. My thanks also go out to Annabel Merullo for first encouraging me to attempt the project, and to the fantastic team at Orion Books and Chronicle Prism who made it possible. Their ideas and encouragement helped improve the book every step of the way.

Several friends and family played essential roles in discussing ideas and checking in on my progress. The Pentons, the Penton Aldersons, the Coles, the Booths, and the Banissys: Your support and interest kept me going. Thanks to Sue for the thought-provoking conversations that shaped my thinking about the importance of touch.

Much of this book was written in the coffee shops of Hitchin: The team at Fussey & Baer deserve a special mention for their daily willingness to discuss the book with me and their cameo appearance in Chapter 7.

Lastly, Mum, Dad, Jasmine, Jamie, and Tegan—writing this book repeatedly reminded me how lucky I've been to have such a supportive group of loved ones to turn to throughout my life. Without you all, I would not be where I am today. Thank you for believing in me and for every moment we've shared. I dedicate this book to you all.

ENDNOTES

Introduction

1. Holt-Lunstad, J., Smith, T. B., Layton, J. B., "Social relationships and mortality risk: a meta-analytic review," *PLOS Med*, 7 (2010): e1000316, DOI: 10.1371/journal.pmed.1000316; Holt-Lunstad, J., "Why Social Relationships Are Important for Physical Health: A Systems Approach to Understanding and Modifying Risk and Protection," *Annu. Rev. Psychol.*, 69 (2018): 437–458, DOI: 10.1146/annurev-psych-122216-011902; Holt-Lunstad, J., Smith, T. B., Baker, M., Harris, T., Stephenson, D., "Loneliness and social isolation as risk factors for mortality: a meta-analytic review," *Perspect. Psychol. Sci.*, 10 (2015): 227–37, DOI: 10.1177/1745691614568352.
2. https://www.cdc.gov/museum/timeline/covid19.html; https://covid19.who.int/
3. https://www.thetimes.co.uk/article/how-britons-have-lost-touch-with-being-tactile-5qqzm5j6x
4. For reading around #MeToo, I recommend *She Said* by Jodi Kantor and Megan Twohey, and *The Women's Atlas* by Joni Seager.
5. Based on analysis of the total sample from the Touch Test for BBC Radio 4 in October 2020.

1. Developmental Touch

1. Bremner, A. J., "The development of touch perception and body representation," in J. J. Lockman and C. S. Tamis-LeMonda (ed.), *The Cambridge Handbook of Infant Development: Brain, behavior, and cultural context* (Cambridge University Press, 2020), 238–62; Bremner, A. J., Spence, C., "The Development of Tactile Perception," in Janette B. Benson (ed.), *Advances in Child Development and Behavior*, 52 (2017): 227–68; Croy, I., Fairhurst, M. T., McGlone, F., "The role of C-tactile nerve fibers in human social development," *Current Opinion in Behavioral Sciences*, 43 (2022): 20–26; Marx, V., Nagy, E., "Fetal behavioral responses to the touch of the mother's abdomen: A frame-by-frame analysis," *Infant Behavior and Development*, 47 (2017): 83–91.

2. Mercuri, M., Stack, D. M., Trojan, S., Giusti, L., Morandi, F., Mantis, I., Montirosso, R., "Mothers' and fathers' early tactile contact behaviors during triadic and dyadic parent–infant interactions immediately after birth and at 3 months postpartum: Implications for early care behaviors and intervention," *Infant Behavior and Development*, 57 (2019): 101347.
3. Bytomski, A., Ritschel, G., Bierling, A., Bendas, J., Weidner, K., Croy, I., "Maternal stroking is a fine-tuned mechanism relating to C-tactile afferent activation: An exploratory study," *Psychology & Neuroscience*, 13(2) (2020): 149–57; Croy, I., Luong, A., Triscoli, C., Hofmann, E., Olausson, H., Sailer, U., "Interpersonal stroking touch is targeted to C tactile afferent activation," *Behav. Brain Res.*, 297 (2016): 37–40, DOI: 10.1016/j.bbr.2015.09.038; Van Puyvelde, M., Collette, L., Gorissen, A. S., Pattyn, N., McGlone, F., "Infants' Autonomic Cardio- Respiratory Responses to Nurturing Stroking Touch Delivered by the Mother or the Father," *Front. Physiol.*, 10 (2019): 1117, DOI: 10.3389/fphys.2019.01117; Van Puyvelde, M., Gorissen, A. S., Pattyn, N., McGlone, F., "Does touch matter? The impact of stroking versus non-stroking maternal touch on cardio-respiratory processes in mothers and infants," *Physiol. Behav.*, 207 (2019): 55–63, DOI: 10.1016/j.physbeh.2019.04.024; Croy, I., Bierling, A., Sailer, U., Ackerley, R., "Individual Variability of Pleasantness Ratings to Stroking Touch Over Different Velocities," *Neuroscience*, 464 (2021): 33–43, DOI: 10.1016/j.neuroscience.2020.03.030.
4. Gursul, D., Goksan, S., Hartley, C., Mellado, G. S., Moultrie, F., Hoskin, A., . . . Slater, R., "Stroking modulates noxious-evoked brain activity in human infants," *Current Biology*, 28(24) (2018): R1380–R1381.
5. Addabbo, M., Licht, V., Turati, C., "Past and present experiences with maternal touch affect infants' attention toward emotional faces," *Infant Behavior and Development*, 63 (2021): 101558.
6. Narvaez, D., Wang, L., Cheng, A., Gleason, T. R., Woodbury, R., Kurth, A., Lefever, J. B., "The importance of early life touch for psychosocial and moral development," *Psicologia: Reflexão e Crítica*, 32 (2019).
7. Ulmer Yaniv, A., Salomon, R., Waidergoren, S., Shimon-Raz, O., Djalovski, A., Feldman, R., "Synchronous caregiving from birth to adulthood tunes humans' social brain," *Proceedings of the National Academy of Sciences*, 118(14) (2021): e2012900118.
8. For an excellent review of work on primates, see Dunbar, R. I., "The social role of touch in humans and primates: behavioral function and neurobiological mechanisms," *Neuroscience & Biobehavioral Reviews*, 34(2) (2010): 260–8.
9. Ionio, C., Ciuffo, G., Landoni, M., "Parent–infant skin-to-skin contact and stress regulation: A systematic review of the literature," *International Journal of Environmental Research and Public Health*, 18(9) (2021): 4695; Field, T., "Pediatric massage therapy research: a narrative review," *Children*, 6(6) (2019): 78; Field, T., Diego, M., Hernandez-Reif, M., "Preterm infant massage therapy research: a review," *Infant Behavior and Development*, 33(2) (2010): 115–24; https://www.unicef.org.uk/babyfriendly/news-and-research/baby-friendly-research/research-supporting-breastfeeding/skin-to-skin-contact/
10. https://www.who.int/publications/i/item/9789241508988; https://www.who.int/news-room/fact-sheets/detail/preterm-birth; Vogel, J. P., Oladapo, O. T., Manu, A., Gulmezoglu, A. M., Bahl, R., "New WHO recommendations to improve the outcomes of preterm birth," *The Lancet Global Health*, 3(10) (2015): e589–e590.

11. Moore, S. R., McEwen, L. M., Quirt, J., Morin, A., Mah, S. M., Barr, R. G., ... Kobor, M. S., "Epigenetic correlates of neonatal contact in humans," *Development and Psychopathology*, 29(5) (2017): 1517–38; Wigley, I. L. C. M., Mascheroni, E., Bonichini, S., Montirosso, R., "Epigenetic protection: maternal touch and DNA-methylation in early life," *Current Opinion in Behavioral Sciences*, 43 (2022): 111–17.
12. https://www.huggies.com/en-ca/why-huggies/about-us/no-baby-unhugged
13. https://www.unicef.org.uk/babyfriendly/about/standards/
14. https://www.unicef.org.uk/babyfriendly/about/standards/
15. Huang, X., Chen, L., Zhang, L. "Effects of paternal skin-to-skin contact in newborns and fathers after cesarean delivery," *The Journal of Perinatal & Neonatal Nursing*, 33(1) (2019): 68–73.
16. https://extension.psu.edu/programs/betterkidcare/early-care/tip-pages/all/touch-why-we-need-it
17. https://www.zerotothree.org/resources/2987-high-five-or-hug-teaching-toddlers-about-consent

2. Scientists Who Stroke

1. Zotterman, Y., "Touch, pain and tickling: an electro-physiological investigation on cutaneous sensory nerves," *The Journal of Physiology*, 95(1) (1939): 1.
2. Nordin, M., "Low-threshold mechanoreceptive and nociceptive units with unmyelinated (C) fibers in the human supraorbital nerve," *The Journal of Physiology*, 426(1) (1990): 229–40.
3. Cascio, C. J., Moore, D., McGlone, F., "Social touch and human development," *Developmental Cognitive Neuroscience*, 35 (2019), 5–11; Loken, L. S., Wessberg, J., McGlone, F., Olausson, H., "Coding of pleasant touch by unmyelinated afferents in humans," *Nature Neuroscience*, 12(5) (2009): 547–8; Croy, I., Sehlstedt, I., Wasling, H. B., Ackerley, R., Olausson, H., "Gentle touch perception: From early childhood to adolescence," *Developmental Cognitive Neuroscience*, 35 (2019): 81–6.
4. Sehlstedt, I., Ignell, H., Backlund Wasling, H., Ackerley, R., Olausson, H., Croy, I., "Gentle touch perception across the life span," *Psychology and Aging*, 31(2) (2016): 176.
5. Olson, W., Dong, P., Fleming, M., Luo, W., "The specification and wiring of mammalian cutaneous low-threshold mechanoreceptors," *Wiley Interdisciplinary Reviews: Developmental Biology*, 5(3) (2016): 389–404.
6. McGlone, F., Reilly, D., "The cutaneous sensory system," *Neuroscience & Biobehavioral Reviews*, 34(2) (2010): 148–59.
7. Mancini, F., Bauleo, A., Cole, J., Lui, F., Porro, A. C., Haggard, P., Iannetti, G. D., "Whole-body mapping of spatial acuity for pain and touch," *Annals of Neurology*, 75, no. 6 (2014): 917–24.
8. Grant, A. C., Thiagarajah, M. C., Sathian, K., "Tactile perception in blind Braille readers: a psychophysical study of acuity and hyperacuity using gratings and dot patterns," *Perception & Psychophysics*, 62(2) (2000): 301–12; Wong, M., Gnanakumaran, V., Goldreich, D., "Tactile spatial acuity enhancement in blindness: evidence for experience-dependent mechanisms," *Journal of Neuroscience*, 31(19) (2011): 7028–37; Sathian, K., Stilla, R., "Cross-modal plasticity of tactile perception in blindness," *Restorative Neurology and Neuroscience*, 28(2) (2010), 271–81; Sadato, N., Pascual-Leone, A., Grafman, J., Ibanez, V., Deiber, M. P., Dold, G., Hallett,

M., "Activation of the primary visual cortex by Braille reading in blind subjects," *Nature*, 380(6574) (1996): 526–8; Burton, H., "Visual cortex activity in early and late blind people," *Journal of Neuroscience*, 23(10) (2003): 4005–11.
9. Tuulari, J. J., Scheinin, N. M., Lehtola, S., Merisaari, H., Saunavaara, J., Parkkola, R., . . . Bjornsdotter, M., "Neural correlates of gentle skin stroking in early infancy," *Developmental Cognitive Neuroscience*, 35 (2019): 36–41.
10. Nieuwenhuys, R., "The insular cortex: a review," *Progress in brain research*, 195 (2012): 123–63.
11. Rolls, E. T., Cheng, W., Feng, J., "The orbitofrontal cortex: reward, emotion and depression," *Brain Commun.*, 16 (2020): fcaa196, DOI: 10.1093/braincomms/fcaa196; Berridge, K. C., Robinson, T. E., "What is the role of dopamine in reward: hedonic impact, reward learning, or incentive salience?", *Brain Res. Rev.*, 28 (1998): 309–69, DOI: 10.1016/s0165-0173(98)00019-8; Cox, J., Witten, I. B., "Striatal circuits for reward learning and decision-making," *Nat. Rev. Neurosci.*, 20 (2019): 482–94, DOI: 10.1038/s41583-019-0189-2; Hyman, S. E., Malenka, R. C., Nestler, E. J., "Neural mechanisms of addiction: the role of reward-related learning and memory," *Annu. Rev. Neurosci.* 29 (2006): 565–98. DOI: 10.1146/annurev.neuro.29.051605.113009.
12. Sailer, U., Triscoli, C., Haggblad, G., Hamilton, P., Olausson, H., Croy, I., "Temporal dynamics of brain activation during 40 minutes of pleasant touch," *Neuroimage* 139 (2016): 360–7, DOI: 10.1016/j.neuroimage.2016.06.031.
13. Walker, S. C., Trotter, P. D., Swaney, W. T., Marshall, A., McGlone, F. P., "C-tactile afferents: Cutaneous mediators of oxytocin release during affiliative tactile interactions?", *Neuropeptides*, 64 (2017): 27–38, DOI: 10.1016/j.npep.2017.01.001; https://www.yourhormones.info/hormones/oxytocin/
14. Aguirre, M., Couderc, A., Epinat-Duclos, J., & Mascaro, O. "Infants discriminate the source of social touch at stroking speeds eliciting maximal firing rates in CT fibers," *Developmental Cognitive Neuroscience*, 36 (2019): 100639.
15. Ravaja, N., Harjunen, V., Ahmed, I., Jacucci, G., Spape, M. M., "Feeling Touched: Emotional Modulation of Somatosensory Potentials to Interpersonal Touch," *Sci. Rep.*, 7 (2017): 40504, DOI: 10.1038/srep40504; Croy, I., Angelo, S. D., Olausson, H., "Reduced pleasant touch appraisal in the presence of a disgusting odor," *PLOS ONE*, 9 (2014): e92975, DOI: 10.1371/journal.pone.0092975; Ellingsen, D. M., Wessberg, J., Chelnokova, O., Olausson, H., Laeng, B., Leknes, S., "In touch with your emotions: oxytocin and touch change social impressions while others' facial expressions can alter touch," *Psychoneuroendocrinology*, 39 (2014): 11–20, DOI: 10.1016/j.psyneuen.2013.09.017; Gazzola, V., Spezio, M. L., Etzel, J. A., Castelli, F., Adolphs, R., Keysers, C., "Primary somatosensory cortex discriminates affective significance in social touch," *Proc. Natl Acad. Sci. USA*, 109 (2012): E1657–66, DOI: 10.1073/pnas.1113211109.
16. Rahman, J., Mumin, J., Fakhruddin, B., "How Frequently Do We Touch Facial T-Zone: A Systematic Review," *Ann. Glob. Health*, 86(1)(2020): 75, DOI: 10.5334/aogh.2956.
17. Blakemore S. J., Wolpert, D., Frith, C. "Why can't you tickle yourself?", *Neuroreport*, 11 (2000): R11-6. doi: 10.1097/00001756-200008030-00002; Press, C., Kok, P., Yon, D. "The perceptual prediction paradox," *Trends in Cognitive Sciences*, 24(1) (2020): 13–24.

18. Yim, J. "Therapeutic benefits of laughter in mental health: a theoretical review," *The Tohoku Journal of Experimental Medicine*, 239(3) (2016): 243-249; Hall, G. S., Alliń, A. "The psychology of tickling, laughing, and the comic," *The American Journal of Psychology*, 9(1) (1897): 1–41.
19. Elvitigala, D. S., Boldu, R., Nanayakkara, S., Matthies, D. J., "TickleFoot: Design, Development and Evaluation of a Novel Foot-Tickling Mechanism That Can Evoke Laughter," *ACM Transactions on Computer-Human Interaction*, 29(3) (2022): 1–23.
20. Mohiyeddini, Changiz, Semple, Stuart, "Displacement behavior regulates the experience of stress in men," Stress, 16:2 (2013), 163–71; Dreisoerner, A., Junker, N. M., Schlotz, W., Heimrich, J., Bloemeke, S., Ditzen, B., van Dick, R., "Self-soothing touch and being hugged reduce cortisol responses to stress: A randomized controlled trial on stress, physical touch, and social identity," *Comprehensive Psychoneuroendocrinology*, 8 (2021), 100091.
21. Murphy, J., Brewer, R., Catmur, C., Bird, G. "Interoception and psychopathology: A developmental neuroscience perspective," *Developmental Cognitive Neuroscience*, 23 (2017): 45–56; Brewer, R., Murphy, J., Bird, G. "Atypical interoception as a common risk factor for psychopathology: A review," *Neuroscience & Biobehavioral Reviews*, 130 (2021): 470–508; Murphy, J., Catmur, C., Bird, G., "Classifying individual differences in interoception: Implications for the measurement of interoceptive awareness," *Psychon. Bull. Rev.*, 26 (2019): 1467–71, DOI: 10.3758/s13423-019-01632-7.
22. Matiz, A., Guzzon, D., Crescentini, C., Paschetto, A., Fabbro, F., "The role of self body brushing vs mindfulness meditation on interoceptive awareness: A non-randomized pilot study on healthy participants with possible implications for body image disturbances," *European Journal of Integrative Medicine*, 37 (2020), 10116.

3. Healthy Touch

1. https://en.wikipedia.org/wiki/Free_Hugs_Campaign
2. https://www.oprah.com/oprahshow/the-gift-of-giving-back/all
3. Packheiser, J., Malek, I. M., Reichart, J. S. et al. "The Association of Embracing with Daily Mood and General Life Satisfaction: An Ecological Momentary Assessment Study," *J Nonverbal Behav*, (2022). https://doi.org/10.1007/s10919-022-00411-8
4. Dueren, A. L., Vafeiadou, A., Edgar, C., & Banissy, M. J. "The influence of duration, arm crossing style, gender, and emotional closeness on hugging behavior," *Acta psychologica*, 221 (2021): 103441; Floyd, K. "All touches are not created equal: Effects of form and duration on observers' interpretations of an embrace," *Journal of Nonverbal Behavior*, 23(4) (1999): 283–299.
5. Clipman, J. M., "A hug a day keeps the blues away: The effect of daily hugs on subjective well-being in college students." Paper presented at the annual meeting of the Eastern Psychological Association, Boston, MA (March 1999).
6. Bavishi, A., Slade, M. D., Levy, B. R., "A chapter a day: Association of book reading with longevity," *Soc. Sci. Med.*, 164 (2016): 44–8, DOI: 10.1016/j.socscimed.2016.07.014.
7. Cohen, S., Janicki-Deverts, D., Turner, R. B., Doyle, W. J., "Does hugging provide stress-buffering social support? A study of susceptibility to upper respiratory infection and illness," *Psychol. Sci.*, 26 (2015):135–47, DOI: 10.1177/0956797614559284.

8. Light, K. C., Grewen, K. M., Amico, J. A., "More frequent partner hugs and higher oxytocin levels are linked to lower blood pressure and heart rate in premenopausal women," *Biological Psychiatry*, 69(1) (2005), 5–21.
9. Cohen, S., Janicki-Deverts, D., Turner, R. B., Doyle, W. J., "Does hugging provide stress-buffering social support? A study of susceptibility to upper respiratory infection and illness," *Psychol. Sci.*, 26 (2015): 135–47, DOI: 10.1177/0956797614559284.
10. Cohen, S., Janicki-Deverts, D., Turner, R. B., Doyle, W. J. "Does hugging provide stress-buffering social support? A study of susceptibility to upper respiratory infection and illness," *Psychological Science*, 26(2) (2015): 135–147.
11. van Raalte, Lisa J., Floyd, Kory, "Daily Hugging Predicts Lower Levels of Two Proinflammatory Cytokines," *Western Journal of Communication*, 85:4 (2021), 487–506, DOI: 10.1080/10570314.2020.1850851.
12. https://www.apa.org/monitor/2010/05/weird; Henrich, J., Heine, S., Norenzayan, A., "The weirdest people in the world?", *Behavioral and Brain Sciences*, 33(2–3) (2010), 61–83, DOI:10.1017/S0140525X0999152X.
13. Ditzen, B., Neumann, I. D., Bodenmann, G., von Dawans, B., Turner, R. A., Ehlert, U., Heinrichs, M., "Effects of different kinds of couple interaction on cortisol and heart rate responses to stress in women," *Psychoneuroendocrinology*, 32 (2007): 565–74, DOI: 10.1016/j.psyneuen.2007.03.011.
14. Jakubiak, Brett K., "Providing support is easier done than said: Support providers' perceptions of touch and verbal support provision requests," *Journal of Experimental Social Psychology*, 96 (2021), 104168, https://doi.org/10.1016/j.jesp.2021.104168.
15. Coan, J. A., Schaefer, H. S., Davidson, R. J., "Lending a hand: Social regulation of the neural response to threat," *Psychological Science*, 17(12) (2006), 1032–9.
16. Kim, B. H., Kang, H. Y., Choi, E. Y., "Effects of handholding and providing information on anxiety in patients undergoing percutaneous vertebroplasty," *J. Clin. Nurs.*, 24 (2015):3459–68, DOI: 10.1111/jocn.12928.
17. Quinn, J. F., "Therapeutic touch as energy exchange: testing the theory," *Adv. Nurs. Sci.*, 6 (1984): 42–9, DOI: 10.1097/00012272-198401000-00007.
18. Soares, M., Oliveira, R., Ros, A., et al., "Tactile stimulation lowers stress in fish," *Nat. Commun.*, 2 (2011), 534, https://doi.org/10.1038/ncomms1547; Separate research in other forms of fish has found slightly different results; for instance, in territorial fish, touch can reduce aggression but not stress responses.
19. Soares, M. C., Oliveira, R. F., Ros, A. F., Grutter, A. S., & Bshary, R. "Tactile stimulation lowers stress in fish," *Nature Communications*, 2(1) (2011): 1–5.
20. Dagnino-Subiabre, A., "Resilience to stress and social touch," *Curr. Opin. Behav. Sci.*, 43 (2022): 75–9, DOI: 10.1016/j.cobeha.2021.08.011
21. Robinson, J., Biley, F. C., Dolk, H., "Therapeutic touch for anxiety disorders," *Cochrane Database of Systematic Reviews*, 3 (2007), Art. No. CD006240, DOI: 10.1002/14651858.CD006240.pub2.
22. Field T., "Massage therapy research review," *Complement. Ther. Clin. Pract.*, 20 (2014): 224–9, DOI: 10.1016/j.ctcp.2014.07.002; Field, T., Diego, M., Hernandez-Reif, M., "Preterm infant massage therapy research: a review," *Infant Behav. Dev.*, 33 (2010):115–24, DOI: 10.1016/j.infbeh.2009.12.004.
23. Naruse, S. M., Moss, M. "Positive massage: An intervention for couples' wellbeing in a touch-deprived era," *European Journal of Investigation in Health, Psychology and Education*, 11(2) (2021): 450–467.

24. Field, T. M., Hernandez-Reif, M., Quintino, O., Schanberg, S., Kuhn, C., "Elder Retired Volunteers Benefit From Giving Massage Therapy to Infants," *J. Appl. Gerontol.*, 17 (1998): 229–39.
25. https://www.mayoclinic.org/diseases-conditions/lymphedema/symptoms-causes/syc-20374682
26. Kutner, J. S., Smith, M. C., Corbin, L., Hemphill, L., Benton, K., Mellis, B. K., et al., "Massage therapy versus simple touch to improve pain and mood in patients with advanced cancer: A randomized trial," *Annals of Internal Medicine*, 149 (2008:138–42; Post-White, J., Fitzgerald, M., Savik, K., Hooke, M. C., Hannahan, A. B., Sencer, S. F., "Massage therapy for children with cancer," *Journal of Pediatric Oncology Nursing*, 26 (2009): 16–28.
27. Field, T., Diego, M., Hernandez-Reif, M., "Massage therapy research," *Developmental Review*, 27 (2007): 75–89; Diego, M. A., Hernandez-Reif, M., Field, T., Friedman, L., Shaw, K., "HIV adolescents show improved immune function following massage therapy," *International Journal of Neuroscience*, 106 (2001): 35–45; Hernandez-Reif, M., Field, T., Ironson, G., Beutler, J., Vera, Y., Hurley, J., et al., "Natural killer cells and lymphocytes are increased in women with breast cancer following massage therapy," *International Journal of Neuroscience*, 115 (2005): 495–510.

4. Touch Hunger

1. https://www.theatlantic.com/magazine/archive/2020/07/can-an-unloved-childlearn-to-love/612253/; Sonuga-Barke, E. J., Kennedy, M., Kumsta, R., Knights, N., Golm, D., Rutter, M., . . . Kreppner, J., "Child-to-adult neurodevelopmental and mental health trajectories after early life deprivation: the young adult follow-up of the longitudinal English and Romanian Adoptees study," *The Lancet*, 389 (2017), 1539–48.
2. https://www.psychologicalscience.org/publications/observer/obsonline/harlows-classic-studies-revealed-the-importance-of-maternal-contact.html
3. Prescott, James W., "Body Pleasure and the Origins of Violence," *Bulletin of the Atomic Scientists*, 31:9 (1975), 10–20.
4. Field, T., "American adolescents touch each other less and are more aggressive toward their peers as compared with French adolescents," *Adolescence*, 34 (1999): 753–8.
5. Floyd, K., "Relational and Health Correlates of Affection Deprivation," *Western Journal of Communication*, 78:4 (2014), 383–403, DOI: 10.1080/10570314.2014.927071.
6. Floyd, K. *The Loneliness Cure: Six Strategies for Finding Real Connections in Your Life.* USA, Adams Media, 2015.
7. Floyd, K. *The Loneliness Cure: Six Strategies for Finding Real Connections in Your Life.* USA, Adams Media, 2015.
8. https://www.bbc.com/future/article/20201006-why-touch-matters-more-than-ever-in-the-time-of-covid-19
9. Field, T., Poling, S., Mines, S., Bendell, D., Veazey, C. "Touch deprivation and exercise during the COVID-19 lockdown April 2020," *Medical Research Archives*, 8(8) (2020).
10. Meijer, L. L., Hasenack, B., Kamps, J. C. C., et al., "Affective touch perception and longing for touch during the COVID-19 pandemic," *Sci. Rep.*, 12 (2022), 3887, https://doi.org/10.1038/s41598-022-07213-4.

11. Von Mohr, M., Kirsch, L. P., Fotopoulou, A., "Social touch deprivation during COVID-19: effects on psychological well-being and craving interpersonal touch," *Royal Society Open Science*, 8(9) (2021), 210287.
12. https://www.verywellmind.com/attachment-styles-2795344
13. Read, S., Comas-Herrera, A., Grundy, E., "Social isolation and memory decline in later-life," *The Journals of Gerontology: Series B*, 75(2) (2020), 367–76.
14. https://www.campaigntoendloneliness.org/the-facts-on-loneliness/
15. https://theconversation.com/we-asked-70-000-people-how-coronavirus-affected-them-what-they-told-us-revealed-a-lot-about-inequality-in-the-uk-143718; Sampogna, G., Giallonardo, V., Del Vecchio, V., Luciano, M., Albert, U., Carmassi, C., . . . Fiorillo, A., "Loneliness in Young Adults During the First Wave of COVID-19 Lockdown: Results From the Multicentric COMET Study," *Front. Psychiatry*, 12 (2021):788139, DOI: 10.3389/fpsyt.2021.788139; Groarke, J. M., Berry, E., Graham-Wisener, L., McKenna-Plumley, P. E., McGlinchey, E., Armor, C., "Loneliness in the UK during the COVID-19 pandemic: Cross-sectional results from the COVID-19 Psychological Well-being Study," *PLOS One*, 15 (2020): e0239698, DOI: 10.1371/journal.pone.0239698; Savage, R. D., Wu, W., Li, J., et al., "Loneliness among older adults in the community during COVID-19: a cross-sectional survey in Canada," *BMJ Open*, 11 (2021):e044517, DOI: 10.1136/bmjopen-2020-044517; https://www.covidsocialstudy.org/results.
16. https://www.cdc.gov/mmwr/volumes/69/wr/mm6932a1.htm?s_cid=mm6932a1_w
17. https://www.bbc.co.uk/sounds/series/m000n484 ; https://www.bbc.co.uk/news/stories-54373924
18. https://www.verywellhealth.com/virtual-cuddle-therapy-5202513
19. https://www.huffingtonpost.co.uk/2015/02/06/mcvities-cuddle-cafe-sweeetest-spot_n_6628992.html
20. https://time.com/3456630/anti-loneliness-chair-japan/
21. https://qoobo.info/index-en/
22. Haynes, A. C., Lywood, A., Crowe, E. M., Fielding, J. L., Rossiter, J. M., Kent, C., "A calming hug: Design and validation of a tactile aid to ease anxiety," *PLOS ONE* 17(3) (2022): e0259838, https://doi.org/10.1371/journal.pone.0259838.
23. Pendry, P., Vandagriff, J. L., "Animal Visitation Program (AVP) Reduces Cortisol Levels of University Students: A Randomized Controlled Trial," *AERA Open*, April 2019, DOI:10.1177/2332858419852592.
24. Keysers, C., Kaas, J. H., Gazzola, V., "Somatosensation in social perception," *Nat. Rev. Neurosci.*, 11 (2010):417–28, DOI: 10.1038/nrn2833.
25. https://baltic.art/uploads/For_All_I_Care_Episode_2_Transcript.pdf

5. Tactile Intimacy

1. Jankowiak, W. R., Volsche, S. L., Garcia, J. R., "Is the romantic–sexual kiss a near human universal?," *American Anthropologist*, 117(3) (2015), 535–9.
2. https://www.npr.org/sections/health-shots/2014/11/17/364054843/whats-in-his-kiss-80-million-bacteria?t=1659365871346
3. Maloney, J. M., Chapman, M. D., Sicherer, S. H., "Peanut allergen exposure through saliva: assessment and interventions to reduce exposure," *J. Allergy Clin. Immunol.*, 118 (2006): 719–24, DOI: 10.1016/j.jaci.2006.05.017.

4. https://www.wsj.com/articles/BL-IRTB-27097; https://www.scientificamerican.com/article/bonobo-sex-and-society-2006-06/
5. https://www.bbc.com/future/article/20210813-the-reasons-humans-started-kissing
6. Floyd, Kory, Boren, Justin P., Hannawa, Annegret F., Hesse, Colin, McEwan, Breanna, Veksler, Alice E., "Kissing in Marital and Cohabiting Relationships: Effects on Blood Lipids, Stress, and Relationship Satisfaction," Western Journal of Communication, 73 (2009): 113–33.
7. Gulledge, Andrew K., Gulledge, Michelle H., Stahmannn, Robert F., "Romantic Physical Affection Types and Relationship Satisfaction," *The American Journal of Family Therapy*, 31:4 (2003), 233–42.
8. Wlodarski, R., Dunbar, R. I., "What's in a kiss? The effect of romantic kissing on mate desirability," *Evolutionary Psychology*, 12(1) (2014), 147470491401200114.
9. Hughes, S. M., Harrison, M. A., Gallup Jr, G. G., "Sex differences in romantic kissing among college students: An evolutionary perspective," *Evolutionary Psychology*, 5(3) (2007), 147470490700500310.
10. Wedekind, C., Seebeck, T., Bettens, F., Paepke, A. J., "MHC-dependent mate preferences in humans," *Proc. R. Soc. Lond. B.*, 260 (1995): 245–9; Wedekind, C., Furi, S., "Body odor preference in men and women: do they aim for specific MHC combinations or simply heterozygosity?," *Proc. R. Soc. Lond. B.*, 264 (1997): 1471–9; Ober, C., Weitkamp, L. R., Cox, N., Dytch, H., Kostyu, D., Elias, S., "HLA and mate choice in humans," *Am. J. Hum. Gen.*, 61 (1997): 497–504; Penn, D. J., "The scent of genetic compatibility: sexual selection and the major histocompatiblity complex," *Ethology*, 801 (2002): 1–21; Thornhill, Randy, Gangestad, Steven W., Miller, Robert, Scheyd, Glenn, McCollough, Julie K., Franklin, Melissa, "Major histocompatibility complex genes, symmetry, and body scent attractiveness in men and women," *Behavioral Ecology*, 14 (2003): 668–78, https://doi.org/10.1093/beheco/arg043.
11. Olofsson, J. K., Nordin, S., "Gender differences in chemosensory perception and event-related potentials," *Chem. Senses*, 29 (2004):629–37, DOI: 10.1093/chemse/bjh066.
12. Linden, D. J. (2016). *Touch: The science of the hand, heart, and mind*. Penguin Books.
13. Metz, Michael E., McCarthy, Barry W., "The "Good-Enough Sex" model for couple sexual satisfaction," *Sexual and Relationship Therapy*, 22:3 (2007), 351–62, DOI: 10.1080/14681990601013492.
14. Curtis, Yvette, Eddy, Lisabeth, Ashdown, Brien K., Feder, Holly, Lower, Timothy, "Prelude to a coitus: Sexual initiation cues among heterosexual married couples," *Sexual and Relationship Therapy*, 27:4 (2012), 322–34, DOI: 10.1080/14681994.2012.734604.
15. Komisaruk, B. R., Wise, N., Frangos, E., Liu, W. C., Allen, K., Brody, S., "Women's clitoris, vagina, and cervix mapped on the sensory cortex: fMRI evidence," *J. Sex. Med.*, 8 (2011):2822–30, DOI: 10.1111/j.1743-6109.2011.02388.x.
16. Wise, N. J., Frangos, E., Komisaruk, B. R., "Brain Activity Unique to Orgasm in Women: An fMRI Analysis," *J. Sex. Med.*, 14 (2017): 1380–91, DOI: 10.1016/j.jsxm.2017.08.014; Komisaruk, B. R., Whipple, B., "Functional MRI of the brain during orgasm in women," *Annu. Rev. Sex Res.*, 16 (2005): 62–86; https://www.vox.com/2015/4/1/8325483/orgasms-science.

17. https://www.verywellmind.com/psychological-benefits-of-an-orgasm-5235580
18. Cox, T. (2020). *Great Sex Starts at 50: How to age-proof your libido*. Murdoch Books.
19. Matthias, R. E., Lubben, J. E., Atchison, K. A., Schweitzer, S. O., "Sexual activity and satisfaction among very old adults: results from a community-dwelling Medicare population survey," *Gerontologist*, 37 (1997): 6–14.
20. Skałacka, K., Gerymski, R., "Sexual activity and life satisfaction in older adults," *Psychogeriatrics*, 19 (2019):195–201, DOI: 10.1111/psyg.12381.
21. Debrot, A., Meuwly, N., Muise, A., Impett, E. A., Schoebi, D., "More Than Just Sex: Affection Mediates the Association Between Sexual Activity and Well-Being," *Pers. Soc. Psychol. Bull.*, 43 (2017): 287–99, DOI: 10.1177/0146167216684124.
22. Burleson, M. H., Trevathan, W. R., Todd, M., "In the mood for love or vice versa? Exploring the relations among sexual activity, physical affection, affect, and stress in the daily lives of mid-aged women," *Archives of Sexual Behavior*, 36 (2007): 357–68; Burgoon, J. K., "Relational message interpretations of touch, conversational distance, and posture," *Journal of Nonverbal Behavior*, 15 (1991): 233–59.

6. Touchy-Feely or Avoid at All Costs

1. Trotter, P. D., McGlone, F., Reniers, R. L. E. P., Deakin, J. F. W., "Construction and validation of the touch experiences and attitudes questionnaire (TEAQ): a self-report measure to determine attitudes toward and experiences of positive touch," *Journal of Nonverbal Behavior*, 42 (2018): 379–416.
2. The Touch Personas task has also not yet been scientifically validated in published research. I've developed them for this book to help us think about how we approach touch from day to day.
3. Krahe, C., Drabek, M. M., Paloyelis, Y., Fotopoulou, A., "Affective touch and attachment style modulate pain: a laser-evoked potentials study," *Philos. Trans. R. Soc. Lond. B. Biol. Sci.*, 371 (2016):20160009, DOI: 10.1098/rstb.2016.0009; von Mohr, M., Krahe, C., Beck, B., Fotopoulou, A., "The social buffering of pain by affective touch: a laser-evoked potential study in romantic couples," *Soc. Cogn. Affect. Neurosci.*, 13 (2018): 1121–1130, DOI: 10.1093/scan/nsy085.
4. Wagner, S. A., Mattson, R. E., Davila, J., Johnson, M. D., Cameron, N. M., "Touch me just enough: The intersection of adult attachment, intimate touch, and marital satisfaction," *Journal of Social and Personal Relationships*, 37 (2020): 1945–67, DOI:10.1177/0265407520910791; Jakubiak, B. K., Fuentes, J. D., Feeney, B. C., "Affectionate Touch Promotes Shared Positive Activities," *Pers. Soc. Psychol. Bull.*, (2022):1461672221083764, DOI: 10.1177/0146167222108376, epub ahead of print, PMID: 35440257.
5. Debrot, A., Stellar, J. E., MacDonald, G., Keltner, D., Impett, E. A., "Is Touch in Romantic Relationships Universally Beneficial for Psychological Well-Being? The Role of Attachment Avoidance," *Pers. Soc. Psychol. Bull.*, 47 (2021):1495–1509, DOI: 10.1177/0146167220977709.
6. Allport, G. W., & Odbert, H. S. "Trait-names: A psycho-lexical study," *Psychological Monographs*, 47(1) (1936): i–171.
7. Digman, J. M. "Personality structure: Emergence of the five-factor model," *Annual Review of Psychology*, 41(1) (1990): 417-440.

8. Soldz, S., & Vaillant, G. E. "The Big Five personality traits and the life course: A 45-year longitudinal study," Journal of Research in Personality, 33 (1999): 208-232.
9. Dorros, S., Hanzal, A., Segrin, C., "The Big Five personality traits and perceptions of touch to intimate and nonintimate body regions," *Journal of Research in Personality*, 42(4) (2008): 1067–73.
10. Wilt, J., Revelle, W., "Extraversion," in M. R. Leary and R. H. Hoyle (ed.), *Handbook of Individual Differences in Social Behavior* (The Guilford Press, 2009), 27–45; Schaefer, M., Heinze, H. J., Rotte, M., "Touch and personality: extraversion predicts somatosensory brain response," *Neuroimage*, 62(1) (2012): 432–8, DOI: 10.1016/j.neuroimage.2012.05.004.
11. Vafeiadou, A., Bowling, N. C., Hammond, C., Banissy, M. J. "Assessing individual differences in attitudes towards touch in treatment settings: Introducing the Touch & Health Scale," *Health Psychology Open*. (2022).
12. Singer, J., "Why can't you be normal for once in your life?" From a "problem with no name" to the emergence of a new category of difference," in M. Corker and S. French (ed.), *Disability Discourse*, Buckingham: Open University Press.
13. Lundqvist, L. O., "Hyper-responsiveness to touch mediates social dysfunction in adults with autism spectrum disorders," *Research in Autism Spectrum Disorders*, 9 (2015): 13–20; Kern, J. K., Trivedi, M. H., Grannemann, B. D., Garver, C. R., Johnson, D. G., Andrews, A. A., . . . Schroeder, J. L., "Sensory correlations in autism," *Autism*, 11(2) (2007), 123–34; Thye, M. D., Bednarz, H. M., Herringshaw, A. J., Sartin, E. B., Kana, R. K., "The impact of atypical sensory processing on social impairments in autism spectrum disorder," *Dev. Cogn. Neurosci.*, 29 (2018): 151–67, DOI: 10.1016/j.dcn.2017.04.010; Mikkelsen, M., Wodka, E. L., Mostofsky, S. H., Puts, N. A., "Autism spectrum disorder in the scope of tactile processing," *Developmental Cognitive Neuroscience*, 29 (2018): 140–50; Baranek, G. T., David, F. J., Poe, M. D., Stone, W. L., Watson, L. R., "Sensory Experiences Questionnaire: discriminating sensory features in young children with autism, developmental delays, and typical development," *J. Child Psychol. Psychiatry*, 47 (2006): 591–601, DOI: 10.1111/j.1469-7610.2005.01546.x.
14. For more on autistic adults from the Touch Test see: Penton, T., Bowling, N., Vafeiadou, A., Hammond, C., Bird, G., & Banissy, M. J. "Attitudes to interpersonal touch in the workplace in autistic and non-autistic groups," *Journal of Autism and Developmental Disorders*, (2022): 1–13.
15. Ward J., "Synesthesia," *Annu. Rev. Psychol.*, 64 (2013):49–75, DOI: 10.1146/annurev-psych-113011-143840.
16. Salinas, J. (2017). *Mirror Touch: Notes from a Doctor Who Can Feel Your Pain.* HarperCollins Publishers.
17. https://aeon.co/essays/neuro-quirks-and-super-human-perceptions
18. Ward, J., Banissy, M. J., "Explaining mirror-touch synesthesia," *Cogn. Neurosci.*, 6 (2015): 118–33, DOI: 10.1080/17588928.2015.1042444; https://theconversation.com/some-people-with-synesthesia-feel-other-peoples-sensations-of-touch-painfuland-pleasurable-96150.
19. Poerio, G. L., Blakey, E., Hostler, T. J., Veltri, T., "More than a feeling: Autonomous sensory meridian response (ASMR) is characterized by reliable changes in affect and physiology," *PLOS ONE*, 13 (2018):e0196645, DOI: 10.1371/journal.pone.0196645; Swart, T. R., Banissy, M. J., Hein, T. P., Bruna, R., Pereda,

E., Bhattacharya, J., "ASMR amplifies low frequency and reduces high frequency oscillations," *Cortex*, 149 (2022): 85–100, DOI: 10.1016/j.cortex.2022.01.004; Swart, T. R., Bowling, N. C., Banissy, M. J., "ASMR-Experience Questionnaire (AEQ): A data-driven step towards accurately classifying ASMR responders," *Br. J. Psychol.*, 113 (2022): 68–83, DOI: 10.1111/bjop.12516; Fredborg, B. K., Champagne-Jorgensen, K., Desroches, A. S., Smith, S. D., "An electroencephalographic examination of the autonomous sensory meridian response (ASMR)," *Conscious Cogn.*, 87 (2021):103053, DOI: 10.1016/j.concog.2020.103053; Smith, S. D., Fredborg, B. K., Kornelsen, J., "Functional connectivity associated with five different categories of Autonomous Sensory Meridian Response (ASMR) triggers," *Conscious Cogn.*, 85 (2020):103021, DOI: 10.1016/j.concog.2020.103021.

7. Touch Culture

1. Jourard, S. M., "An exploratory study of body accessibility," *British Journal of Social and Clinical Psychology*, 5 (1966): 221–31; Jourard, S. M., *Disclosing man to himself*, Van Nostrand, 1968.
2. Dutton, J., Johnson, A., Hickson, M., "Touch revisited: observations and methodological recommendations," *Journal of Mass Communication & Journalism*, 7(5) (2017).
3. Suvilehto, J. T., Glerean, E., Dunbar, R. I., Hari, R., Nummenmaa. L., "Topography of social touching depends on emotional bonds between humans," *Proc. Natl Acad. Sci. USA*, 112 (2015): 13811–6, DOI: 10.1073/pnas.1519231112.
4. Barnlund, D. C., *Public and private self in Japan and the United States. Communicative styles of two cultures*, Tokyo, Japan: Simul Press (1975); Dibiase, R., Gunnoe, J., "Gender and culture differences in touching behavior," *The Journal of Social Psychology*, 144(1) (2004): 49–62; Remland, M. S., Jones, T. S., Brinkman, H., "Interpersonal distance, body orientation, and touch: Effects of culture, gender, and age," *The Journal of Social Psychology*, 135(3) (1995): 281–97.
5. Suvilehto, J. T., Nummenmaa, L., Harada, T., Dunbar, R. I. M., Hari, R., Turner, R., Sadato, N., Kitada, R., "Cross-cultural similarity in relationship-specific social touching," *Proc. Biol. Sci.*, 286 (2019): 20190467, DOI: 10.1098/rspb.2019.0467.
6. Sorokowska, A., Sorokowski, P., Hilpert, P., Cantarero, K., Frackowiak, T., Ahmadi, K., . . . Pierce Jr, J. D., "Preferred interpersonal distances: a global comparison," *Journal of Cross-Cultural Psychology*, 48(4) (2017): 577–92.
7. IJzerman, H., Semin, G. R., "Temperature perceptions as a ground for social proximity," *Journal of Experimental Social Psychology*, 46 (2010): 867–73; Zhong, C. B., Leonardelli, G. J., "Cold and lonely: does social exclusion literally feel cold?," *Psychological Science*, 19 (2008): 838–42; Williams, L. E., Bargh, J. A., "Experiencing physical warmth promotes interpersonal warmth," Science, 322 (2008): 606–7; Sorokowski, P., Sorokowska, A., Onyishi, I. E., Szarota, P., "Montesquieu hypothesis and football: Players from hot countries are more expressive after scoring a goal," *Polish Psychological Bulletin*, 44 (2013): 421–30.
8. Sorokowska, A., Saluja, S., Sorokowski, P., Frąckowiak, T., Karwowski, M., Aavik, T., . . . Croy, I., "Affective interpersonal touch in close relationships: a cross-cultural perspective," *Personality and Social Psychology Bulletin*, 47(12) (2021): 1705–21; Schaller, M., Murray, D. R., "Pathogens, personality, and culture: Disease prevalence predicts worldwide variability in sociosexuality, extraversion, and openness

to experience," *Journal of Personality and Social Psychology*, 95 (2008): 212–21; Sorokowska, A., Sorokowski, P., Hilpert, P., Cantarero, K., Frackowiak, T., Ahmadi, K., . . . Pierce Jr, J. D., "Preferred interpersonal distances: a global comparison," *Journal of Cross-Cultural Psychology*, 48(4) (2017): 577–92.

9. Carney, D. R., Jost, J. T., Gosling, S. D., Potter, J., "The secret lives of liberals and conservatives: Personality profiles, interaction styles, and the things they leave behind," *Political Psychology*, 29(6) (2008): 807–40, https://doi.org/10.1111/j.1467-9221.2008.00668.x; Klofstad, C. A., McDermott, R., Hatemi, P. K., "The dating preferences of liberals and conservatives," *Political Behavior*, 35(3) (2013): 519–38, https://doi.org/10.1007/s11109-012-9207-z; Tybur, J. M., Inbar, Y., Aaroe, L., Barclay, P., Barlow, F. K., de Barra, M., Žeželj, I., "Parasite stress and pathogen avoidance relate to distinct dimensions of political ideology across 30 nations," *Proceedings of the National Academy of Sciences of the United States of America*, 113(44) (2016): 12408–13, https://doi. org/10.1073/pnas.1607398113.

10. Sorokowska, A., Sorokowski, P., Hilpert, P., Cantarero, K., Frackowiak, T., Ahmadi, K., . . . Pierce Jr, J. D., "Preferred interpersonal distances: a global comparison," *Journal of Cross-Cultural Psychology*, 48(4) (2017): 577–92; Rapp, M. A., Gutzmann, H., "Invasions of personal space in demented and nondemented elderly persons," *International Psychogeriatrics*, 12 (2000): 345–52; Hall, E. T., *The Hidden Dimension*, New York, NY: Doubleday, 1966; Webb, J. D., Weber, M. J., "Influence of sensory abilities on the interpersonal distance of the elderly," *Environment & Behavior*, 35 (2003): 695–711; Ozdemir, A., "Shopping malls: Measuring interpersonal distance under changing conditions and across cultures," *Field Methods*, 20 (2008): 226–48.

11. Burleson, M. H., Roberts, N. A., Coon, D. W., & Soto, J. A. (2019). Perceived cultural acceptability and comfort with affectionate touch: Differences between Mexican Americans and European Americans. Journal of Social and Personal Relationships, 36(3), 1000–1022.

12. Dhawan, E. (2021). *Digital Body Language: How to build trust & connection no matter the distance*. St Martin's Press.

8. Social Touch

1. Kraus, M. W., Huang C., Keltner, D., "Tactile communication, cooperation, and performance: an ethological study of the NBA," *Emotion*, 10 (2010): 745–9, DOI: 10.1037/a0019382.
2. https://www.mirror.co.uk/sport/football/news/jurgen-klopp-urges-liverpools-players-8860127
3. https://www.football365.com/news/klopp-explains-the-importance-of-hugging-players
4. https://www.sportingnews.com/us/nba/news/suns-highfives-winning-percentage-study-research-important/s25dsgvj4uyp1t5ibifcl48tb
5. Milius, I., Gilbert, W. D., Alexander, D., Bloom, G. A., "Coaches' Use of Positive Tactile Communication in Collegiate Basketball," *International Sport Coaching Journal*, 8(1) (2020): 91–100.
6. Crusco, A. H., Wetzel, C. G., "The Midas touch: The effects of interpersonal touch on restaurant tipping," *Personality and Social Psychology Bulletin*, 10(4) (1984): 512–17.

7. Kaufman, Douglas, Mahoney, John M., "The Effect of Waitresses' Touch on Alcohol Consumption in Dyads," *The Journal of Social Psychology*, 139:3 (1999): 261–7, DOI: 10.1080/00224549909598383.
8. Gueguen, N., Jacob, C., "The effect of touch on tipping: an evaluation in a French bar," *International Journal of Hospitality Management*, 24(2) (2005): 295–9.
9. Hornik, J., "Effects of physical contact on customers' shopping time and behavior," *Marketing Letters*, 3(1) (1992): 49–55.
10. Liu, Y., Zang, X., Chen, L., Assumpcao, L., Li, H., "Vicariously touching products through observing others' hand actions increases purchasing intention, and the effect of visual perspective in this process: An fMRI study," *Human Brain Mapping*, 39(1) (2018): 332–43.
11. Luangrath, A. W., Peck, J., Gustafsson, A., "Should I touch the customer? Rethinking interpersonal touch effects from the perspective of the touch initiator," *Journal of Consumer Research*, 47(4) (2020): 588–607.
12. Gueguen, N., Fischer-Lokou, J., "Another evaluation of touch and helping behavior," *Psychological Reports*, 92(1) (2003): 62–4.
13. Kleinke, C. L., "Compliance to requests made by gazing and touching experimenters in field settings," *Journal of Experimental Social Psychology*, 13(3) (1977): 218–23; Joule, R. V., Gueguen, N., "Touch, compliance, and awareness of tactile contact," *Perceptual and Motor Skills*, 104(2) (2007): 581–8.
14. Eaton, M., Mitchell-Bonair, I. L., Friedmann, E., "The effect of touch on nutritional intake of chronic organic brain syndrome patients," *Journal of Gerontology*, 41(5) (1986): 611–16.
15. Nannberg, J. C., Hansen, C. H., "Post-compliance touch: An incentive for task performance," *The Journal of Social Psychology*, 134(3) (1994): 301–7.
16. Hoffmann, L., Kramer, N. C., "The persuasive power of robot touch. Behavioral and evaluative consequences of non-functional touch from a robot," *PLOS ONE*, 16(5) (2021): e0249554.
17. Li, H., Cao, Y., "The Dark Side of Interpersonal Touch: Physical Contact Leads to More Non-compliance With Preventive Measures to COVID-19," *Psychological Reports*, (2021), 00332941211051985.
18. Martin, B. A., "A stranger's touch: Effects of accidental interpersonal touch on consumer evaluations and shopping time," *Journal of Consumer Research*, 39(1) (2012): 174–84.
19. Saarinen, A., Harjunen, V., Jasinskaja-Lahti, I., Jaaskelainen, I. P., Ravaja, N., "Social touch experience in different contexts: A review," *Neuroscience & Biobehavioral Reviews*, 131 (2021): 360–72; Shamloo, S. E., Carnaghi, A., Fantoni, C., "Investigating the relationship between intergroup physical contact and attitudes towards foreigners: the mediating role of quality of intergroup contact," *PeerJ*, 6 (2018): e5680.
20. Camps, J., Tuteleers, C., Stouten, J., Nelissen, J., "A situational touch: How touch affects people's decision behavior," *Social Influence*, 8(4) (2013): 237–50.
21. https://spsp.org/news/character-and-context-blog/klebl-physical-appearance-prejudice
22. Novembre, G., Etzi, R., "Hedonic responses to touch are modulated by the perceived attractiveness of the caresser," *Neuroscience*, 464 (2021): 79–89; Patterson,

M. L., Powell, J. L., Lenihan, M. G., "Touch, compliance, and interpersonal affect," *Journal of Nonverbal Behavior*, 10(1) (1986): 41–50; Burgoon, J. K., Walther, J. B., Baesler, E. J., "Interpretations, evaluations, and consequences of interpersonal touch," *Human Communication Research*, 19(2) (1992): 237–63.

23. Langlois, J. H., Ritter, J. M., Casey, R. J., & Sawin, D. B. Infant attractiveness predicts maternal behaviors and attitudes. *Developmental Psychology*, 31(3) (1995): 464.
24. Boderman, A., Freed, D. W., Kinnucan, M. T., "'Touch Me, Like Me': Testing an Encounter Group Assumption," *The Journal of Applied Behavioral Science*, 8(5) (1972): 527–33.
25. https://www.simplypsychology.org/mere-exposure-effect.html

9. Do Touch, Don't Touch

1. Gutek, B. A., Morasch, B., Cohen, A. G., "Interpreting social-sexual behavior in a work setting," Journal of Vocational Behavior, 32 (1983): 30–48; Dougherty, T. W., Turban, D. B., Olson, D. E., Dwyer, P. D., Lapreze, M. W., "Factors affecting perceptions of workplace sexual harassment," *Journal of Organizational Behavior*, 17 (1996): 489–501.
2. https://www.hrmagazine.co.uk/content/news/workers-want-physical-contact-banned-at-work/
3. Based on an unpublished sub-analysis of data from Penton, T., et al., "Attitudes to Interpersonal Touch in the Workplace in Autistic and non-Autistic Groups," *Journal of Autism and Developmental Disorders* (in press).
4. https://www.huffingtonpost.co.uk/entry/simone-biles-gives-emotional-testimony_n_62a0845ae4b0cf43c8430381
5. Hertenstein, M. J., Keltner, D., App, B., Bulleit, B. A., Jaskolka, A. R., "Touch communicates distinct emotions," *Emotion*, 6(3) (2006): 528; App, B., McIntosh, D. N., Reed, C. L., Hertenstein, M. J., "Nonverbal channel use in communication of emotion: how may depend on why," *Emotion*, 11(3) (2011): 603.
6. https://www.apa.org/news/press/releases/2012/03/well-being
7. Gu, Y., Ocampo, J. M., Algoe, S. B., Oveis, C., "Gratitude expressions improve teammates' cardiovascular stress responses," *Journal of Experimental Psychology: General*, 2022.
8. Marler, L. E., Cox, S. S., Simmering, M. J., Bennett, R. J., Fuller, J. B., "Exploring the role of touch and apologies in forgiveness of workplace offenses," *Journal of Managerial Issues*, (2011): 144–63.
9. Caza, A., Zhang, G., Wang, L., Bai, Y., "How do you really feel? Effect of leaders' perceived emotional sincerity on followers' trust," *The Leadership Quarterly*, 26(4) (2015): 518–31.
10. Fuller, B., Simmering, M. J., Marler, L. E., Cox, S. S., Bennett, R. J., Cheramie, R. A., "Exploring touch as a positive workplace behavior," *Human Relations*, 64(2) (2011): 231–56.
11. https://www.bbc.co.uk/news/world-us-canada-52506079
12. https://www.bbc.com/worklife/article/20200413-coronavirus-will-covid-19-endthe-handshake

13. https://www.ft.com/content/bfe91a50-87fa-43d8-b16d-4c5d124069da
14. Oxlund, B., "An anthropology of the handshake," *Anthropology Now*, 12(1) (2020): 39–44.
15. Stewart, G. L., Dustin, S. L., Barrick, M. R., Darnold, T. C., "Exploring the handshake in employment interviews," *Journal of Applied Psychology*, 93(5) (2008): 1139.
16. Schroeder, J., Risen, J. L., Gino, F., Norton, M. I., "Handshaking promotes deal-making by signaling cooperative intent," *Journal of Personality and Social Psychology*, 116(5) (2019): 743.

10. Digital Touch

1. https://teslasuit.io and https://teslasuit.io/use-case/worlds-first-haptic-rugby-tackle/
2. https://www.economist.com/special-report/2021/04/08/the-rise-of-working-from-home
3. https://www.independent.co.uk/tech/mark-zuckerberg-teleporting-transporting-clubhouse-b1798022.html
4. https://tech.fb.com/ar-vr/2020/05/the-future-of-work-and-the-next-computing-platform/
5. https://innovation.microsoft.com/en-us/tech-minutes-future-vr-haptics
6. https://www.ultraleap.com/haptics/
7. Obrist, M., Subramanian, S., Gatti, E., Long, B., Carter, T., "Emotions mediated through mid-air haptics," *Proceedings of the 33rd Annual ACM Conference on Human Factors in Computing Systems*, April 2015, 2053–62.
8. Culbertson, H., Schorr, S. B., Okamura, A. M., "Haptics: The present and future of artificial touch sensation," *Annual Review of Control, Robotics, and Autonomous Systems*, 1(1) (2018): 385–409.
9. https://www.upi.com/Defense-News/2018/12/26/First-Harris-T7-bomb-disposal-robots-sent-to-British-army/6771545850861/
10. https://www.convergerobotics.com/; https://www.wired.com/story/how-i-became-a-robot-in-london/; https://www.newsweek.com/jeff-bezos-robotics-robot-hands-re-mars-telerobot-haptx-shadow-robot-company-1442556
11. Smids, J., Nyholm, S., Berkers, H., "Robots in the workplace: a threat to—or opportunity for—meaningful work?," *Philosophy & Technology*, 33(3) (2020), 503–22.
12. Cramer, H., Kemper, N., Amin, A., Wielinga, B., Evers, V., "'Give me a hug': the effects of touch and autonomy on people's responses to embodied social agents," *Computer Animation and Virtual Worlds*, 20(2–3) (2009): 437–45.
13. Block, A. E., Kuchenbecker, K. J., "Emotionally supporting humans through robot hugs," *Companion of the 2018 ACM/IEEE International Conference on Human–Robot Interaction*, March 2018, 293–4). For studies on human hug duration see Dueren, A. L., Vafeiadou, A., Edgar, C., & Banissy, M. J. "The influence of duration, arm crossing style, gender, and emotional closeness on hugging behavior," *Acta psychologica*, 221 (2021): 103441.
14. https://techcrunch.com/2015/12/07/olly/; https://consent.yahoo.com/v2/collectConsent?sessionId=3_cc-session_ba850787-685f-4300-bd99-9df15b59f564
15. https://hi.is.mpg.de/research_projects/huggiebot-evolution-of-an-interactive-hugging-robot-with-visual-and-haptic-perception

16. https://www.cwplus.org.uk/
17. https://www.miro-e.com/
18. http://www.parorobots.com/
19. Moyle, W., et al., "Social robots helping people with dementia: Assessing efficacy of social robots in the nursing home environment," *6th International Conference on Human System Interactions* (HSI), 2013, 608–13; Wada, K., Shibata, T., Saito, T., Sakamoto, K., Tanie, K., "Psychological and social effects of one year robot assisted activity on elderly people at a health service facility for the aged," *Proceedings of the 2005 IEEE International Conference on Robotics and Automation*, 2005, 2785–90; Wada, K., Shibata, T., "Living with seal robots—its sociopsychological and physiological influences on the elderly at a care house," *IEEE Transactions on Robotics*, 23 (2007):972–80; Shibata, T., et al., "Mental commit robot and its application to therapy of children," *IEEE/ASME International Conference on Advanced Intelligent Mechatronics*, 2001, 1053–8; Eskander, R., Tewari, K., Osann, K., Shibata, T., "Pilot study of the PARO therapeutic robot demonstrates decreased pain, fatigue, and anxiety among patients with recurrent ovarian carcinoma," *Gynecologic Oncology*, 130 (2013):e144–e145; Trost, M. J., Ford, A. R., Kysh, L., Gold, J. I., Matarić, M., "Socially Assistive Robots for Helping Pediatric Distress and Pain," *The Clinical Journal of Pain*, 35 (2019):451–8.
20. https://us.aibo.com/
21. Banks, M. R., Willoughby, L. M., Banks, W. A., "Animal-assisted therapy and loneliness in nursing homes: use of robotic versus living dogs," *Journal of the American Medical Directors Association*, 9(3) (2008): 173–7.
22. Geva, N., Uzefovsky, F., Levy-Tzedek, S., "Touching the social robot PARO reduces pain perception and salivary oxytocin levels," *Sci. Rep.*, 10 (2020):9814, DOI: 10.1038/s41598-020-66982-y.
23. https://www.microsoft.com/en-us/research/publication/virtual-reality-without-vision-a-haptic-and-auditory-white-cane-to-navigate-complex-virtual-worlds/
24. Novich, S. D., Eagleman, D. M., "Using space and time to encode vibrotactile information: toward an estimate of the skin's achievable throughput," *Experimental Brain Research*, 233(10) (2015): 2777–88.
25. Perrotta, M. V., Asgeirsdottir, T., & Eagleman, D. M. Deciphering sounds through patterns of vibration on the skin. *Neuroscience*. 458 (2021): 77-86.
26. Petrini, F. M., Bumbasirevic, M., Valle, G., Ilic, V., Mijović, P., Čvančara, P., . . . Raspopovic, S., "Sensory feedback restoration in leg amputees improves walking speed, metabolic cost and phantom pain," *Nature Medicine*, 25(9) (2019): 1356–63.
27. https://www.snexplores.org/article/sense-touch-haptics-virtual-reality-prosthetics
28. https://www.digicatapult.org.uk/news-and-insights/publications/post/haptics-what-the-future-feels-like/
29. https://feeldoppel.com/; Azevedo, R., Bennett, N., Bilicki, A., Hooper, J., Markopoulou, F., Tsakiris, M., "The calming effect of a new wearable device during the anticipation of public speech," *Scientific Reports*, 7(1) (2017): 1–7.
30. Yao, H., Yang, W., Cheng, W., Tan, Y. J., See, H. H., Li, S., . . . Tee, B. C., "Near-hysteresis- free soft tactile electronic skins for wearables and reliable machine learning," *Proceedings of the National Academy of Sciences*, 117(41) (2020): 25352–9; Sanderson, K., "Electronic skin: from flexibility to a sense of touch," *Nature*, 591(7851) (2021): 685–7.

31. Yu, X., Xie, Z., Yu, Y., Lee, J., Vazquez-Guardado, A., Luan, H., . . . Rogers, J. A., "Skin-integrated wireless haptic interfaces for virtual and augmented reality," *Nature*, 575(7783) (2019): 473–9.
32. https://xenoma.com/
33. https://cutecircuit.com/hugshirt/
34. This question was first posed to me by Professor Carey Jewitt in our Touch Test Radio Episode on Digital Touch: you can hear more here: https://www.bbc.co.uk/sounds/play/m000n6r1.
35. Guillén, M. (2020). *2030: How Today's Biggest Trends Will Collide and Reshape the Future of Everything*. St Martin's Press.